Preface

How do we grow?
How do we control our actions?
How do we learn about the world around us?

As you read this book, you will discover the answers to these questions – and many more besides. You will be invited to look at yourself and at the way you do things. You will learn about the development of your body, and a little about the development of your mind as well.

This book is concerned with the biology of *ourselves*. It has been produced in conjunction with an exhibition of the same name which opened at the British Museum (Natural History) in May 1977. The exhibition, the first of the Museum's major new exhibition programme, represents an exciting new approach to learning about natural history. Much of the material used in the exhibition is presented here in this book.

Preparing the exhibition was a mammoth task involving the effort and imagination of a great many people, both within the Museum and outside, and I should like to take this opportunity of thanking everyone concerned. In particular, for their invaluable help as advisers to the exhibition I should like to thank Professor J. Z. Young, Wellcome Institute of the History of Medicine, London (general adviser), Professor R. L. Gregory, University of Bristol (perception), Dr A. R. Jonckheere and Margaret Redshaw, University College, London (cognitive psychology), Professor J. M. Tanner and Dr M. Preece, Institute of Child Health, London (endocrinology and growth), and Professor L. Wolpert, Middlesex Hospital (physical development). I should also like to thank the following for their advice and help: Dr M. Argyll, University of Oxford (perception), Dr H. G. Davies and Dr S. Neidle, King's College, London (genetics), Professor J. Frisby, University of Sheffield (perception), Dr J. O'Keefe, University College, London (brain research), Professor R. O'Rahilly, Carnegie Embryological Laboratories, University of California (embryology), Dr J. S. Parker, Queen Mary College (genetics), Professor P. E. Polani and Dr M. G. Daker, Paediatric Research Unit, Guy's Hospital (genetics) and Dr J. Versey and Angela Hobsbaum, Institute of Education, London (cognitive psychology).

R. H. Hedley
Director
British Museum (Natural History)
May 1977

5

Chapter 1

Living cells

What are our bodies made of?

Our bodies are made of millions of tiny living cells. Cells make up our skin, our bones, our muscles and brains, and all the other parts of our bodies. Everything we do involves millions of tiny cells of different shapes and sizes, all working together.

There are more than 50 000 000 000 000 cells in one human body. Most of them are so small that it would take 100 000 of them to cover the head of a pin.

Each cell has its own job to do, but all cells live, grow and finally die.

What do we see if we look inside a cell?

Although our cells are so small, they are usually quite complex. A typical cell has many different parts, each with its own job to do.

The protective barrier
Each cell is completely enclosed by a protective *membrane*. This membrane keeps out harmful substances, but allows useful substances to enter the cell. Waste products and materials made by the cell are allowed to pass out through the membrane.

The energy producers
There are many *mitochondria* inside the cell. Chemical reactions inside the mitochondria produce energy for all the cell's activities. A very active cell usually has lots of mitochondria.

The information centre
Inside the cell, enclosed by its own membrane, is the *nucleus* (plural: nuclei). This is the cell's information centre. It stores all the instructions the cell needs to live and grow and carry out all its activities. These instructions are carried on two sets of threadlike *chromosomes*.

You can find out more about chromosomes in Chapter 3.

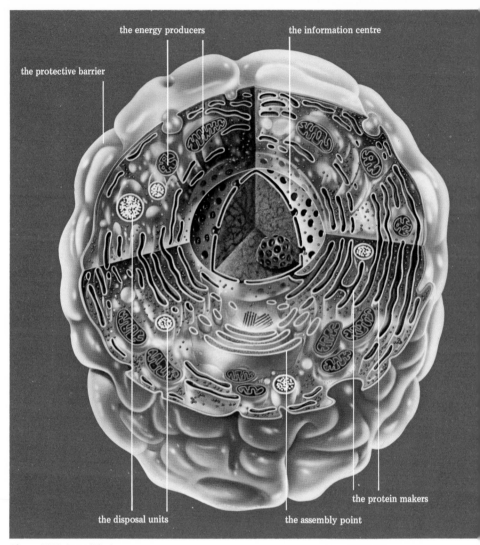

the energy producers

the information centre

the protective barrier

the protein makers

the disposal units

the assembly point

About 5 000 times lifesize

The protein makers
The network of tubes and sacs inside the cell is the *endoplasmic reticulum*, where proteins are made. The beadlike *ribosomes* sticking to the endoplasmic reticulum are also involved in this process. If there are no ribosomes on the endoplasmic reticulum, no proteins are being made there.

The assembly point
All the substances made by the cell are assembled and stored by the *Golgi complex*. Some are then passed out of the cell to be used elsewhere.

The disposal units
The *lysosomes* contain destructive enzymes which digest harmful particles and old or worn-out parts of the cell.

Many different cells . . .

Our bodies are made of many different kinds of cells. Although these cells are all built to the same basic pattern, their shape and contents vary according to the particular job each cell has to do . . .

Food-absorbing cells

When food has been digested, it must be absorbed into the bloodstream so that it can be carried to all parts of the body. Digested food is absorbed by cells lining the small intestine (outlined in white below).

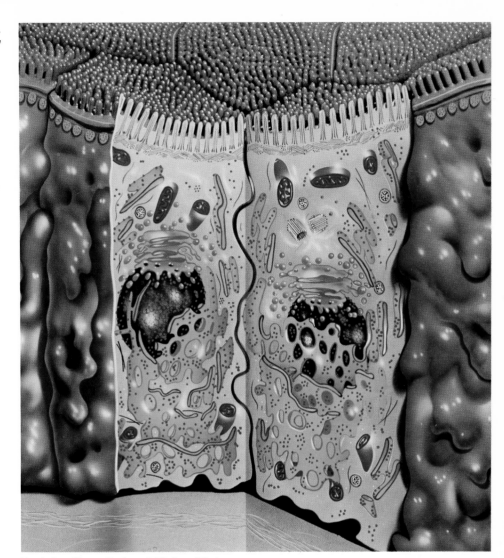

Food-absorbing cell, about 5000 times lifesize

Food-absorbing cells are joined together at the sides, but the membrane on the top surface of each one is free and extremely folded. This folding gives each cell a very large area for absorbing food.

Food is digested outside the cell by enzymes around the folds. It is then absorbed through the membrane into the cell. The absorbed food passes down through the cell and across the lower membrane into the bloodstream.

9

Blood cells

Blood travels round our bodies in a closed system of tubes. The largest of these are our arteries and veins. The smallest are the tiny capillaries, which reach every part of the body.

One drop of blood contains about 5 000 000 red cells and 5000 white cells, all floating in liquid plasma.

Red cells

Red blood cells carry oxygen from the lungs to the rest of the body. Each cell is filled with red haemoglobin and has no other contents, not even a nucleus. In the lungs, oxygen is absorbed through the cell's thin membrane and combines with the haemoglobin inside. The cell is then carried round the body in the bloodstream. It gives up its oxygen to other cells, and is carried back to the lungs for more.

Red cells are among the smallest cells in the body. They are soft and flexible, so they can squeeze through even the smallest capillaries. But they are not very tough and, because they have no nucleus, they do not live very long. So they constantly need to be replaced.

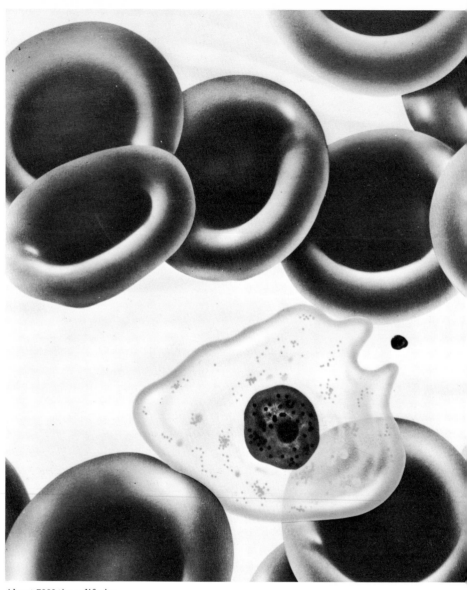

About 7000 times lifesize

White cells

There are several different kinds of white blood cells. The most common type have lots of lysosomes and look very granular. Their main job is to destroy harmful microbes and other unwanted particles.

The white cell changes shape as it moves along. It flows round each particle until it is completely surrounded. Digestive enzymes from the lysosomes then destroy the particle. When its lysosomes are used up, the white cell has to be replaced.

There are two special cells . . .

When a *sperm cell* from a man meets an *ovum* (egg cell) from a woman in the right place at the right time, these two special cells join together to form one new cell. This new cell then grows and divides and gradually develops into a baby.

Sperm cells

Sperm cells are produced in a man's *testes*. During his lifetime, a man produces many many millions of sperm cells. Sperm cells are stored in the *sperm ducts* and, at any time,

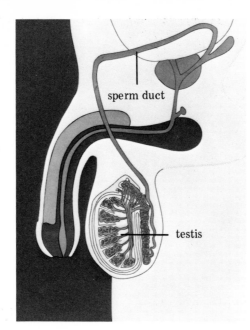

millions of them can be released. They are released in a liquid called seminal fluid.

Sperm cells are among the smallest cells in the body. Each one has a definite head end, a middle piece and a tail. The head end contains the nucleus, which has only one set of chromosomes (the nuclei of other cells contain two sets). In front of the nucleus is a store of digestive enzymes.

The mitochondria in the middle piece provide the sperm cell with energy, and it can swim by lashing its tail.

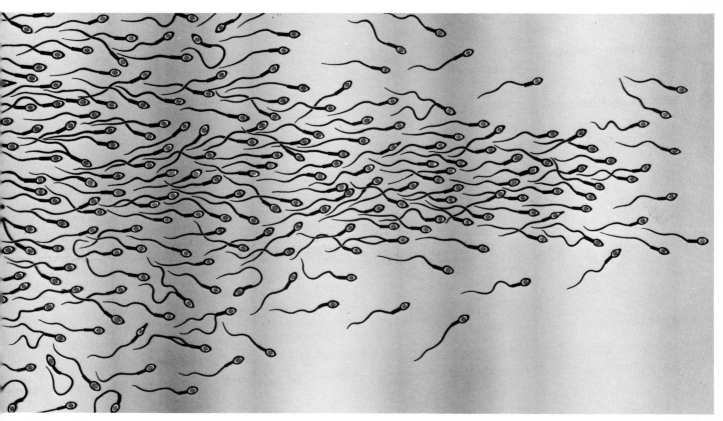

about 500 times lifesize

The ovum

Ova (egg cells) are produced in a woman's *ovaries*. Each month, one ovum becomes ripe and is released from one of the ovaries into a nearby *oviduct*. Tiny 'hairs' inside the oviduct help the ovum drift down towards the *uterus* (womb). The ovum is ripe for about two days. If it does not meet a sperm cell while it is ripe, the ovum dies.

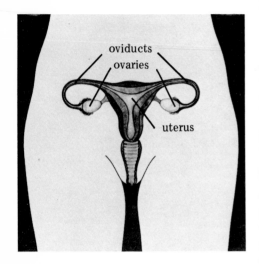

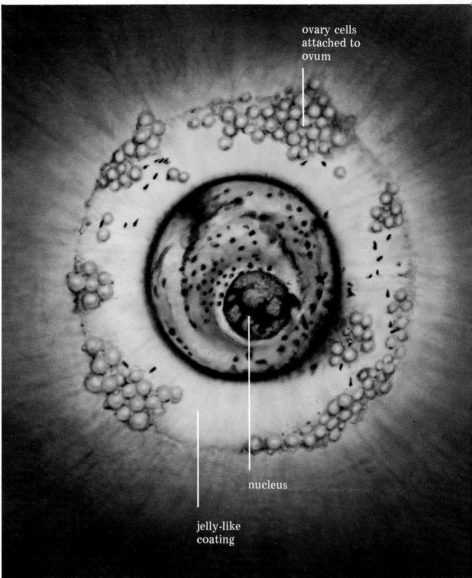

ovary cells attached to ovum

nucleus

jelly-like coating

500 times lifesize

An ovum is larger than any other cell in the human body and can just be seen with the naked eye. Like the nucleus of the sperm cell, the ovum nucleus contains only one set of chromosomes.

The ovum has a jelly-like coating which prevents it from sticking to the sides of the oviduct. Inside the ovum, there are lots of fatty droplets (yolk). These will nourish the ovum for the first few hours if it meets a sperm cell and begins to develop into a baby.

How the special cells meet

Because a baby needs a protected place in which to develop, the sperm cell meets the ovum inside the woman's body. The baby can then develop in the woman's uterus, which has been specially prepared for it.

How sperm cells enter a woman's body

A direct route to the ovum is provided when a man and a woman are linked together during *sexual intercourse*. A small amount of seminal fluid, containing millions of sperm cells, is expelled through the man's penis and deposited just below the woman's uterus. The sperm cells must then make their own way to the ovum in the oviduct.

The journey to the ovum

Helped by movements of the woman's uterus and vagina, some of the sperm cells manage to swim into the uterus. Many get no further. Some are too exhausted. Others are killed by the warmth inside the uterus.

Only the strongest sperm cells pass through the uterus into oviducts. And only about half of these will enter the correct oviduct. Eventually, a few hundred sperm cells may approach the ovum...

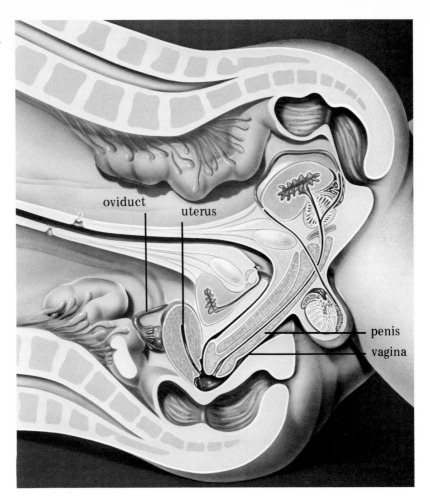

Fertilization

An ovum is *fertilized* when a sperm cell enters it. A few hundred sperm cells may reach the ovum, but only one will actually fertilize it.

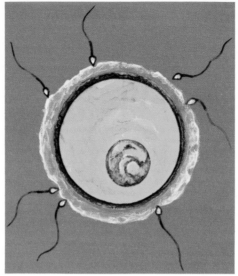

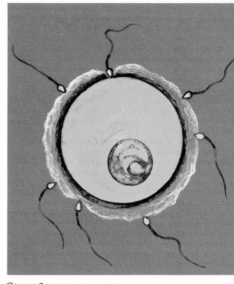

Stage 1
When the sperm cells are near the ovum, they release their stores of digestive enzymes. These enzymes help the sperm cells penetrate the ovum's jelly-like coating.

Stage 2
A few sperm cells pass right through the jelly-like coating and reach the membrane of the ovum.

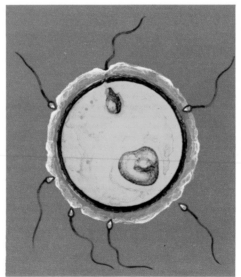

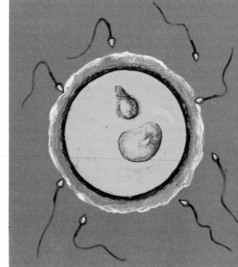

Stage 3
One sperm cell actually penetrates the membrane. Its head and middle piece enter the ovum, and the membrane closes behind them. Changes in the membrane and the jelly-like coating now prevent other sperm cells from entering the ovum.

Stage 4
The nucleus of the sperm cell moves towards the nucleus of the ovum. The membranes around the nuclei break down, and the two nuclei merge into one. The ovum has now been fertilized . . .

Diagrams about 350 times lifesize

The nucleus of the fertilized ovum contains two sets of chromosomes – one set from the father and one set from the mother. In Chapter 3, you can find out why this is so important.

Chapter 2

Growing

The tiny fertilized ovum grows and
divides and gradually develops into
a baby.

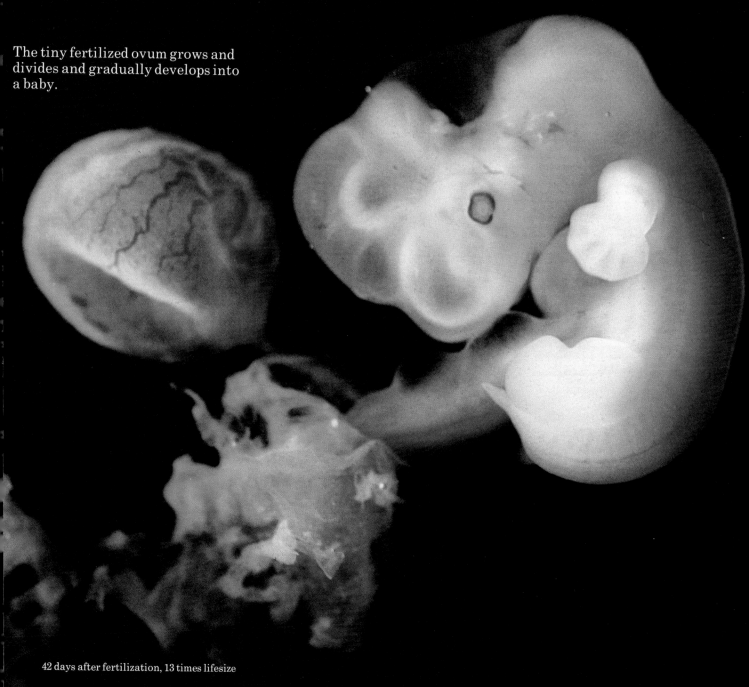

42 days after fertilization, 13 times lifesize

Life before birth

About a day after fertilization, the ovum divides into two. Then, less than a day later, these two cells divide, so there are now four cells. After a few hours, these four cells divide, forming eight cells.

The cells continue to divide, becoming smaller and smaller. By the third day, the ovum has become a ball of tiny dividing cells. It is still protected by the jelly-like coating, and is still being swept along the oviduct towards the uterus.

The ball of cells reaches the uterus three or four days after fertilization. The jelly-like coating then begins to break down and, by the fifth day, it has completely disappeared.

The cells inside the ball are now dividing more slowly than the others, and the ball is made up of two distinct types of cell. The group of cells inside the ball will form the baby. The flattened cells on the outside will form the structures that protect and nourish the baby as it develops.

A week after fertilization, the ball of cells begins to burrow into the wall of the uterus. The outer cells absorb nourishment from the uterus and pass it to the inner cells. This means that, for the first time, the cells can now increase in size before dividing, so the ball starts to grow bigger.

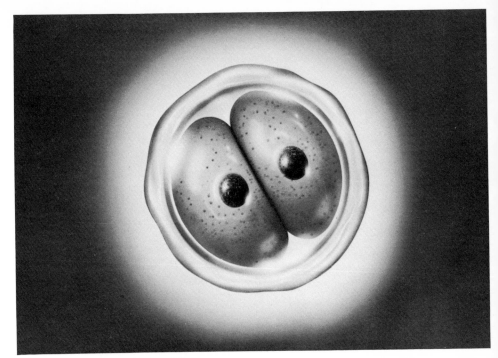

1 day after fertilization, 350 times lifesize

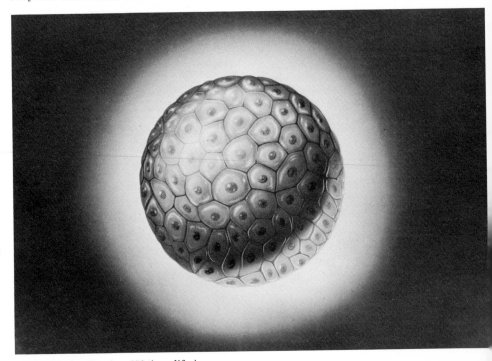

5 days after fertilization, 350 times lifesize

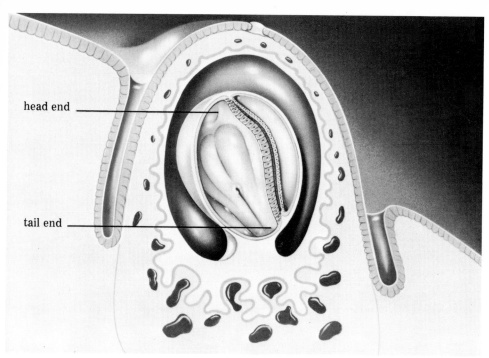

head end

tail end

18 days after fertilization, 30 times lifesize

Inside the ball, the cells of the developing baby have arranged themselves into a disc. As these cells continue to grow and divide and move into position, the disc gradually becomes pear-shaped. The developing baby now has a definite head end, a tail end, a back and a front.

The cells inside the developing baby are very active as they divide and grow and begin to form the brain, heart, gut and spinal cord. Twenty-five days after fertilization, the baby is still no bigger than a grain of rice – yet its developing heart has already started to beat.

The baby develops rapidly as cells begin to form many other parts of its body. Its arms appear as buds on the side of the body, and soon the leg buds appear as well. At this stage, the body looks segmented, with a bent head and a curled tail. But, as the baby continues to develop, the segments disappear.

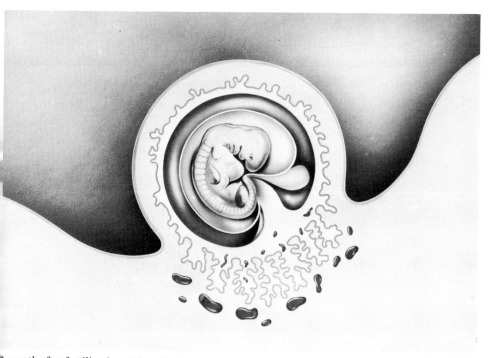

1 month after fertilization, 5 times lifesize

Six weeks after fertilization, the baby's muscles begin to form, and its hands and feet. Gradually the eyes and ears form, the tail shrinks away, and the face begins to look human.

Two and a half months after fertilization, the baby is almost perfectly formed. Even though it is still small enough to fit inside a chicken's egg, it has a heart, brain, digestive system, muscles, cartilage, and even bones. In fact, all the main parts of the body are now formed, though they are not yet fully developed.

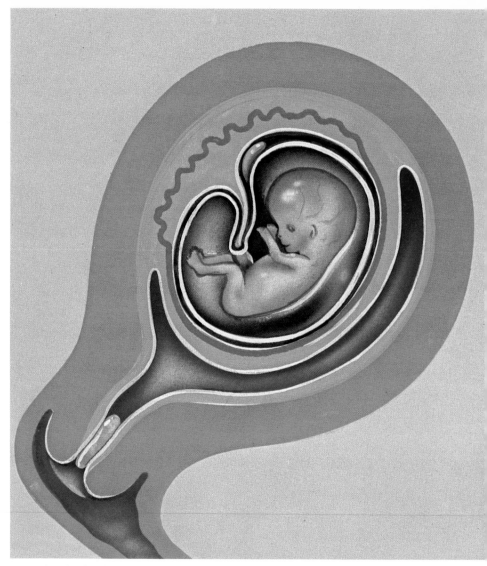

$2\frac{1}{2}$ months after fertilization, lifesize

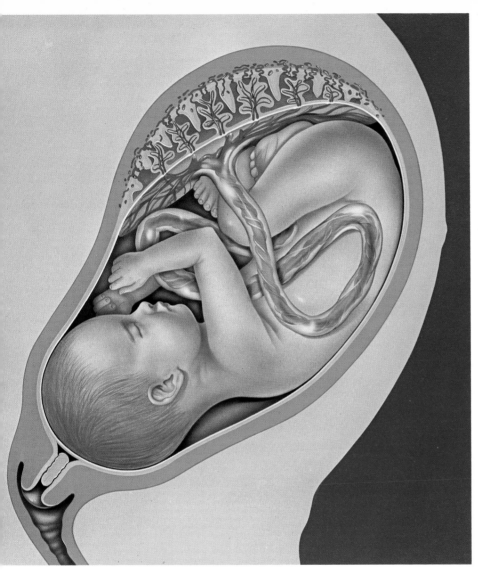

months after fertilization, $\frac{1}{3}$ lifesize

Over the next few months, the baby grows very rapidly and begins to change shape as its limbs increase in length. The mother's abdomen begins to bulge noticeably, and she can feel the baby moving about.

As the baby grows larger, there is less space for it to move about and, about two months before birth, it comes to rest in a head-down position, ready to be born.

Nine months after fertilization, the baby is fully equipped to leave the protection of the uterus and face the world outside its mother.

Birth

Birth is the first major event in the baby's life. It occurs in three stages: *labour, delivery* and *afterbirth*.

You can find out how birth is controlled on page 79.

Labour

1 The baby is resting in a head-down position, ready to be born.

2 The mother feels the contractions of the muscles in the uterus wall. This is the beginning of labour.

3 The contractions gradually get stronger and more frequent, pushing the baby out of the uterus into the vagina.

Delivery

Every two or three minutes, powerful contractions of the uterus force the baby's head further through the vagina. Eventually the baby's head emerges from the vagina into the outside world. This is the most tiring and difficult part of birth but, after only a few more contractions, the baby is born.

Afterbirth

While the baby was in the uterus, it received food and oxygen from its mother via the *placenta*.

A few minutes after birth, contractions of the uterus force the placenta out of the mother's body. (This is why the placenta is usually known as the afterbirth.) Because the baby no longer needs the placenta, it is removed by cutting through the umbilical cord.

The placenta – the unborn baby's life-support system

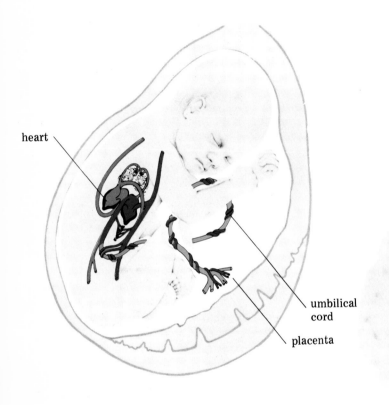

heart

umbilical cord

placenta

lungs

heart

gut

Before birth
The baby's blood flows through the umbilical cord to the placenta, where it receives food and oxygen from the mother's blood.

After birth
The baby breathes in oxygen through its lungs and starts to eat and digest its own food. The baby's blood circulation changes so that more blood flows to the lungs and gut.

Beginnings of independence

The newborn baby's first cries open up the millions of tiny airways to its lungs and, with its first breath, the road to independence has begun.

While the baby was in its mother's uterus, it was warm and protected. It received food and oxygen, and all its wastes were removed. As soon as it is born, the baby must begin to fend for itself.

What does the baby have to do for itself?

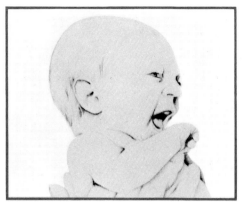

breathe

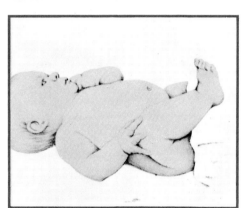

excrete

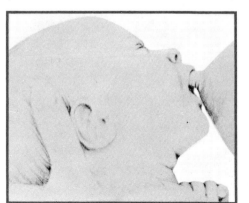

eat

What does the mother still have to do for the baby?

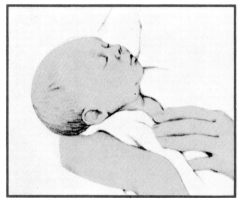

keep it warm

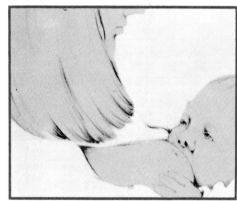

feed it

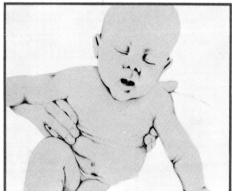

clean it

Growth

As the baby grows into an adult, the number of cells in its body increases. The visible effect of this cell multiplication can conveniently be measured as increases in height or weight. But the baby does not just grow bigger and heavier. Its shape and body proportions also change as it grows up.

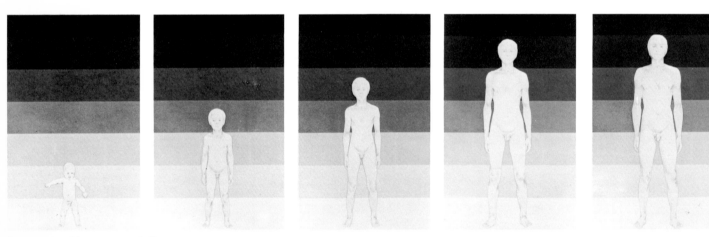

As you grow older, you grow taller.

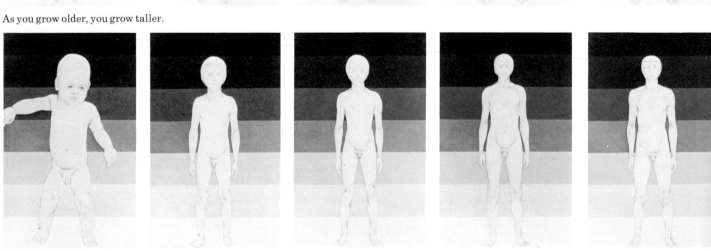

And your shape and body proportions change.

The basis of growth is more cells . . .

When a new cell is about to divide, it takes in nourishment and grows to twice its original size. The membrane around the nucleus disappears, and a furrow starts to form around the cell. The nucleus divides into two, and the furrow round the cell gradually deepens, eventually splitting the cell into two new cells.

About 1000 times lifesize

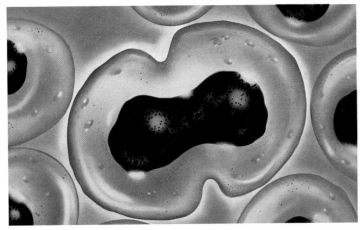

3 Preparing to divide

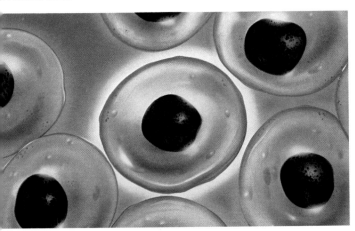

One cell

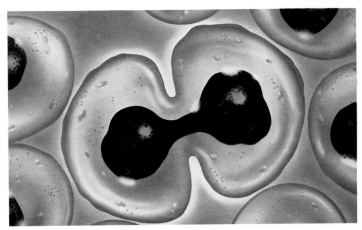

4 Dividing

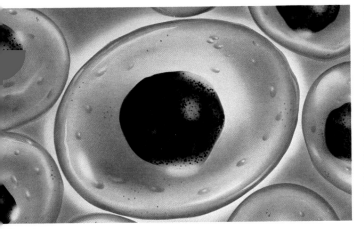

Growing

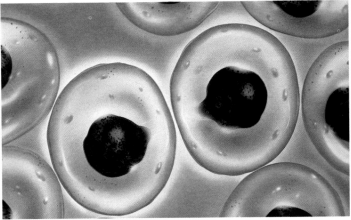

5 Two cells

Even when you have finished growing . . .

Even in an adult, cells in many parts of the body need to grow and divide to replace their dead and dying neighbours. And cells which cannot be replaced need to be kept in working order.

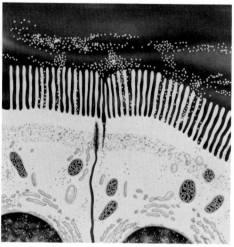

Food-absorbing cells in the gut need to grow and divide to replace the ones damaged by food particles. About 10 000 times lifesize.

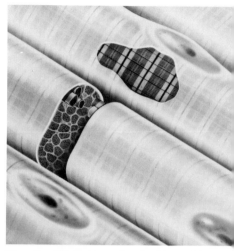

Muscle cells must be maintained in working order. Because they are so specialized, they cannot be replaced. About 1000 times lifesize.

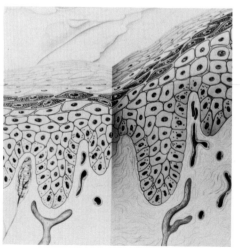

Skin cells must rapidly grow and divide to replace the dead cells continually worn away from the surface. About 1000 times lifesize.

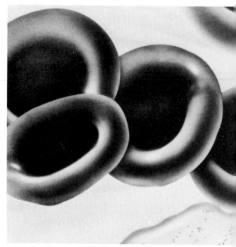

Thousands of new **red blood cells** are needed every second to replace the ones that die. Because red blood cells have no nucleus and cannot divide, they have to be replaced by new cells formed in the bone marrow. About 5000 times lifesize.

chromosomes

We are all different. But we are all like our parents. Why?

Every one of us is a product of
- the environment we grow up in, and
- the genetic instructions we inherit from our parents. Both of these factors are vitally important.

We all inherit different instructions, and we all grow up in different physical and social circumstances.

The instructions

Each of us develops from a single cell, the fertilized ovum. This single cell contains the *instructions* for all the millions of cells that make up the body.

Where do the instructions come from?

Equal quantities of genetic instructions come from the ovum and the sperm cell which fertilized it. The instructions are stored on the threadlike *chromosomes* inside the nuclei of these two cells.

The nucleus of the ovum contains 23 chromosomes, each carrying a different part of the instructions. The sperm cell nucleus also contains 23 chromosomes, carrying a similar set of instructions.

During fertilization, the two nuclei merge into one. So the fertilized ovum contains two sets of genetic instructions – one set from the mother and one set from the father.

What are chromosomes really like?

Most of the time, chromosomes are very long and so fine that they cannot be seen, even with a high-power microscope.

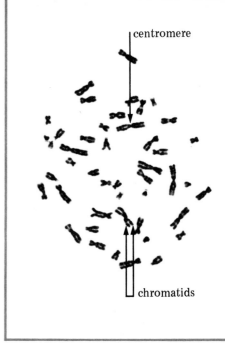

Chromosomes from a dividing human cell 2000 times lifesize

But, before a cell divides, its chromosomes double and coil up very tightly. These coiled-up chromosomes can be seen under the microscope. They look like dark, double rods.

The two parts of each double chromosome are called *chromatids* and the point where they join is called the *centromere*.

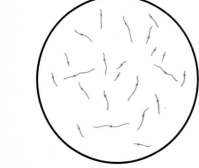

Nucleus of sperm cell (blue) and of ovum

Nucleus of a human cell 2000 times lifesize

What are the instructions made of, and how do they work?

DNA
The genetic instructions are carried by the double-stranded spiral molecule known as *DNA* (short for Deoxyribo-Nucleic-Acid).

Each chromatid probably contains just one, very long, molecule of DNA.

The genetic code
The instructions are in the form of a *code*. This code is made up of four different chemical units (shown here as four different coloured shapes). These units are linked in pairs across the DNA molecule.

(What do you notice about the way the units are paired?)

The order of the paired units along the molecule varies according to what instructions are being coded. A sequence of several hundred paired units is usually needed to code each instruction.

The DNA in each chromosome codes hundreds of different instructions.

Proteins
DNA codes the instructions for making *proteins*. Proteins are large complex molecules that form many of the substances that make up our bodies. They also take part in all the different processes that keep us alive. Insulin, haemoglobin, enzymes, antibodies and hair – these are all proteins.

Our bodies make thousands of different proteins. And at least one genetic instruction is needed for each protein.

Templates
The instructions for making proteins are stored in the cell nucleus. When a particular protein is needed, special *'messenger RNA'* molecules are copied from the instructions and pass out from the nucleus into the surrounding cell. Here the 'messenger RNA' molecules act as *templates* on which the protein molecules are built up.

(Have you noticed how a brown unit is always paired with an orange unit, and a purple with a yellow? – the chemical units in a DNA molecule always pair in this way.)

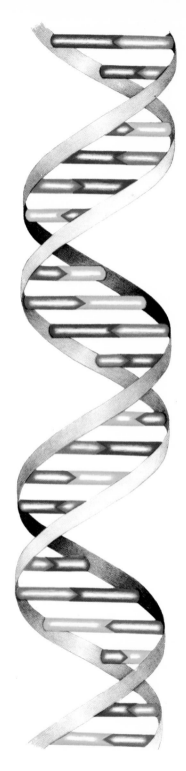

In this diagram, the DNA molecule is shown simply as two spiral strands linked by the paired units.

29

The instructions pass to every cell

The 46 chromosomes in the fertilized ovum carry the instructions for all the cells that make up the body. As the ovum develops, it divides over and over again, and an exact copy of all the instructions is passed to each new cell.

How are the instructions copied?

In these diagrams, the DNA molecule is again shown as two spiral strands linked by the paired units.

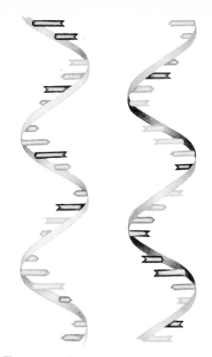

2 The two strands separate.

4 In this way, two new molecules of DNA are formed.

Because the chemical units of the DNA molecule can only pair in a certain way, the two new DNA molecules are identical. They are also exactly the same as the original DNA molecule – and carry exactly the same coded instructions.

The two new DNA molecules carry the instructions for two identical chromosomes. But, at this stage, the two future chromosomes are joined together as the *chromatids* of a double chromosome.

All 46 chromosomes in the cell nucleus double in this way before the cell divides.

1 A double-stranded DNA molecule.

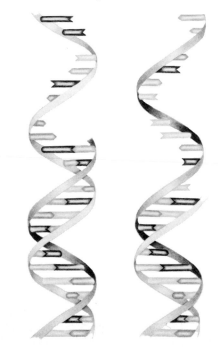

3 A new matching strand is built up beside each strand.

Sharing out the instructions

When the cell divides, the two identical copies of the genetic instructions must be divided equally so that each new cell gets the complete copy of all the instructions. This is done by a special sharing-out process called *mitosis*. Mitosis ensures that one chromatid from each double chromosome passes to each new cell.

For simplicity, these diagrams show only two chromosomes in the nucleus.

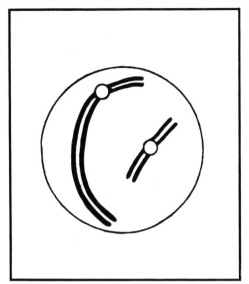

1 Each chromosome appears as two identical *chromatids* joined at a *centromere*.

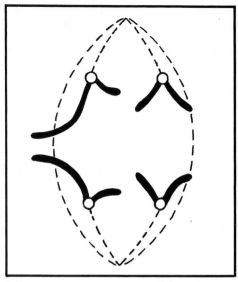

3 Each centromere separates, and the two chromatids of each double chromosome move to opposite ends of the spindle.

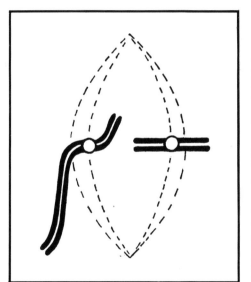

2 The nuclear membrane disappears, and a system of fibres ('the *spindle*') forms. Each centromere attaches to a fibre near the centre or *equator* of the spindle.

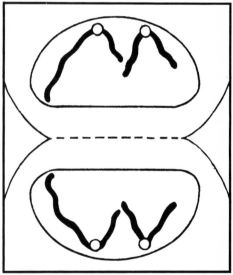

4 Two new nuclei are formed as a new membrane develops around each group of chromatids (now known as chromosomes).

Eventually the whole cell splits into two new cells, each with a nucleus containing 46 chromosomes.

The 46 human chromosomes

These are the chromosomes from a human cell, stained and photographed when the cell was about to divide.

The chromosomes can be paired together according to their shape, size and banding pattern. If you look carefully, you will be able to find an exact partner for each of the 22 numbered chromosomes.

(You may not be able to find an exact partner for the chromosome labelled X – later in the chapter, you will find out why).

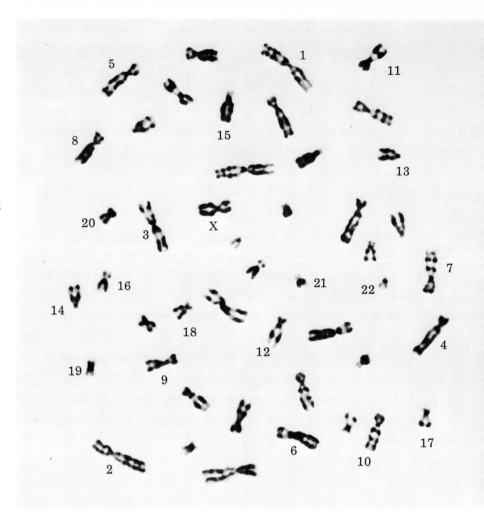

About 3000 times lifesize

There are 23 pairs of chromosomes in the nucleus of every cell. With one possible exception, the two chromosomes in each pair not only look the same, they also carry the same genetic instructions. So there are two sets of instructions in each nucleus. These are exact copies of the two sets of instructions that were in the original fertilized ovum – one set from the mother, and one set from the father.

Alternative instructions

The two sets of instructions are not always identical. For instance, an instruction for eye colour from one parent might be 'make them brown', and the corresponding instruction from the other parent might be 'make them blue'.

Many genetic instructions are like this – they can occur in two, three or even more different forms.

Inheriting the instructions

The sperm cell and the ovum are special cells which each contain only one set of instructions (23 chromosomes). The nuclei of these special cells are formed by a special division process called *meiosis*.

Ova and sperm cells are formed from cells with two sets of instructions (46 chromosomes). Meiosis shuffles the two sets of instructions and then separates them so that every ovum and sperm cell receives its own unique set of genetic instructions.

For simplicity, these diagrams show only two pairs of chromosomes in the nucleus. The maternal chromosomes (from the mother) are shown in red, and the paternal chromosomes (from the father) in blue.

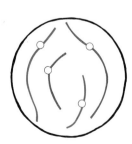

Each chromosome appears as a single thread.

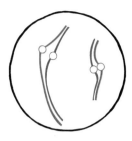

2 The chromosomes arrange themselves into their pairs. Each pair is known as a *bivalent*.

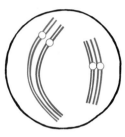

3 The chromosomes double. Each bivalent now consists of two identical maternal and two identical paternal chromatids.

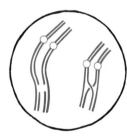

4 Breaks appear in one paternal and one maternal chromatid of each bivalent. The broken ends mend in a criss-cross way or *cross-over*, which recombines the two chromatids.

5 The nuclear membrane disappears and a spindle forms. The two centromeres of each bivalent attach to a spindle fibre and begin to move apart.

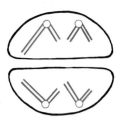

6 The double chromosomes of each bivalent move to opposite ends of the spindle. The spindle disappears, and two new nuclei form.

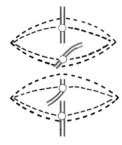

7 The nuclear membranes disappear and a spindle forms around each group of double chromosomes. Each double chromosome attaches to a spindle fibre, and its two chromatids begin to move apart.

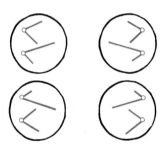

8 Four nuclei are formed as a new membrane develops around each group of chromatids (now known as chromosomes). These will eventually form the nuclei of sperm cells (in a man) or ova (in a woman).

Why we are all different

Meiosis produces four nuclei which each contain 23 different chromosomes. These 23 chromosomes correspond to the 23 *pairs* of chromosomes in the original cell. So each of the four nuclei contains one set of instructions. And each of these four sets of instructions represents a different combination of maternal and paternal instructions.

One reason for this is the cross-overs. During meiosis, at least one cross-over occurs at some point along every bivalent. Because a cross-over recombines two chromatids, it also recombines the instructions they carry.

(Turn back to page 33. After crossing over, two of the chromatids of each bivalent are part blue and part red.)

The second reason is that during meiosis, the chromatids sort themselves in a random way. This means that each nucleus may receive *any one* of the four chromatids from each bivalent – it seems to be a matter of chance which chromatid ends up in which nucleus.

As a result of meiosis, every ovum and every sperm cell carries a unique combination of genetic instructions. So every fertilized ovum contains two unique sets of instructions – one from the mother and one from the father.

In this way, sexual reproduction ensures that every one of us (with the exception of identical twins) inherits a different combination of genetic instructions.

34

Boy or girl – what decides?

The sex of a baby is decided by one special pair of chromosomes – the *sex chromosomes*. The nucleus of every cell of a man's body contains one 'X' sex chromosome and one, quite different, 'Y' sex chromosome.

(Turn back to page 32. Are the chromosomes in the photograph from a man or a woman?)

As a result of meiosis, every ovum receives one X sex chromosome. But a sperm cell can receive *either* an X *or* a Y sex chromosome.

The sex of a baby is determined by the sort of sperm cell that fertilizes the ovum. If the ovum is fertilized by a sperm cell carrying an X sex chromosome, it will develop into a girl (XX). If the ovum is fertilized by a sperm cell carrying a Y sex chromosome, it will develop into a boy (XY).

You can find out more about the way the different sexes develop in Chapter 8.

Boy or girl – what are the chances?

A man produces equal numbers of X- carrying and Y-carrying sperm cells. So there is always an *equal* chance that any baby will be a boy or a girl – just as there is always an equal chance that a coin will come up heads or tails. The sex of one child does not influence the sex of the next.

A, B, AB or O – what determines your blood group?

The ABO blood group system is well known, mainly because of its importance in blood transfusions. Your blood group is determined by a genetic instruction which can occur in three slightly different forms – A, B and O. One form codes 'make substance A', the second form codes 'make substance B' and the third form codes 'make (inactive) substance O'.

Each of us has two of these instructions in the nucleus of every cell – one from mother and one from father. There are six possible pairs of these instructions (AA, AB, AO, BB, BO and OO).

But, because substance O is inactive, some of the pairs have the same effect:

AA and AO both produce blood group A,
BB and BO both produce blood group B,
AB produces blood group AB,
OO produces blood group O.
So there are only four different ABO blood groups.

The inheritance of ABO blood groups is much more complicated than the inheritance of sex. Although you inherit your blood group from your parents, your blood group could be different from either of theirs – or different from the blood groups of your brothers and sisters.

For example, suppose a man's chromosomes carry instructions A and O, and his wife's carry instructions B and O. The man will be blood group A, but he will produce equal numbers of A-carrying and O-carrying sperm cells. His wife will be blood group B, but she will produce equal numbers of B-carrying and O-carrying ova. This means that these two people could produce a child with any of the four blood groups. (If you want to check this, look at the chart.)

Suppose these parents have only one child. Can you work out the chances (one in how many?) of that child having any particular blood group?

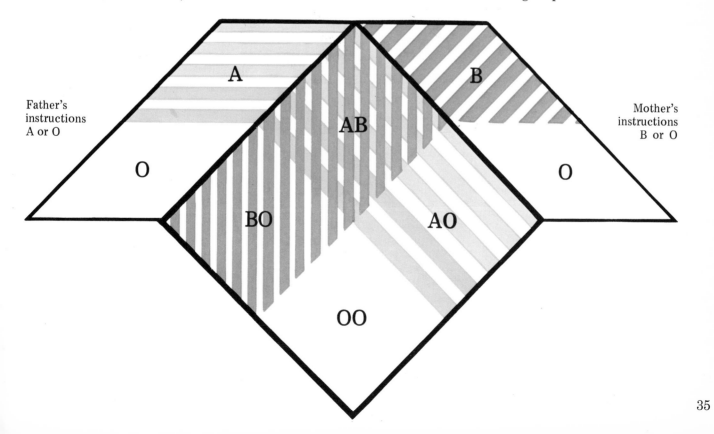

Father's instructions A or O

Mother's instructions B or O

How many instructions?

Nobody knows how many different genetic instructions we have. Estimates generally vary between 10 000 and 100 000 instructions. It is also estimated that, in most people, about 10 per cent of the instructions occur in two different forms (like the AB, AO or BO pairs of instructions for ABO blood groups).

From this sort of information, we can calculate that it is practically impossible for any two people (except for identical twins) to receive exactly the same combination of genetic instructions.

What each of us becomes . . .

The way we develop does not depend on our genetic instructions alone. The food we eat, the air we breathe, the animals and people we live with – these also influence our development. What each of us becomes depends both on our surroundings and on our genetic instructions, and the continual interaction between them.

Chapter 4

Movement

Have you ever wondered how you move?

What moves?
– Bones

Your body is supported and shaped by a framework of *bones*. These act as a system of levers, allowing different parts of your body to move relative to one another.

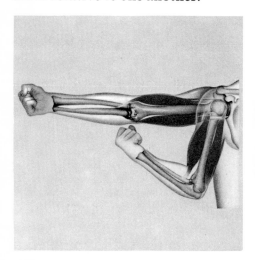

When you bend your arm, the bones in your lower arm move relative to the bones in your upper arm.

Moving bones

In most parts of your body, the bones are not actually joined. Instead, they fit closely together, forming *joints*. At each joint, the bones are linked by tough, flexible *ligaments*.

The joints between your bones allow you to move in different ways. You have many different kinds of joints, and each one allows you to carry out a different sort of movement.

Look carefully at the joints between the bones of the skeleton. Then try moving your own body to see what kinds of movement each joint allows.

In a few parts of your body, the bones are actually joined together, and cannot be moved. For example, the bones of your skull are joined firmly together, providing a strong protective box for your brain.

You can find out more about the brain at the end of the chapter.

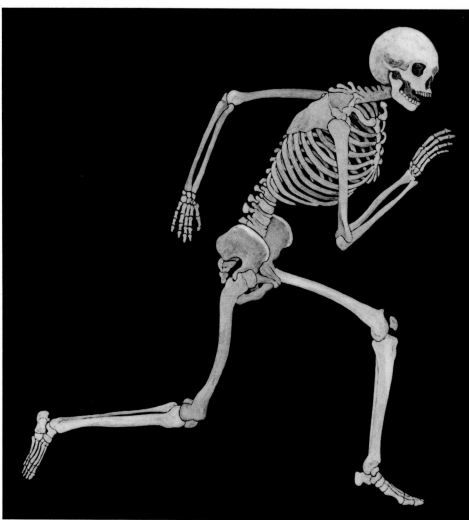

What makes bones move? - Muscles

Bend your arm.
The movement is caused by *muscles* pulling on the bones of your arm.

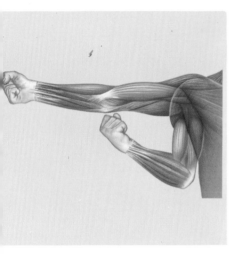

Muscles can only *pull*, they cannot push. This is why most of your muscles are arranged in opposing pairs. When one muscle has pulled a bone in one direction, its partner has to pull it back. (The opposing action of a pair of muscles is an example of *antagonism* – there is another example on page 68.)

Joining muscles to bones

Inside a muscle, there are bundles of long *fibres* (muscle cells) held together by tough connecting sheaths. A similar sheath round the outside binds the whole muscle together.

At each end of the muscle, all these connecting sheaths join together, forming the *tendon* which anchors the muscle to a bone. A tendon is tough and flexible, but cannot be stretched.

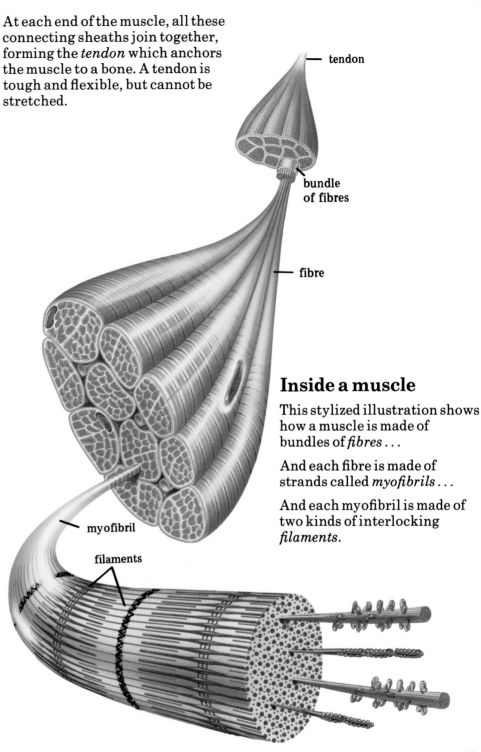

tendon

bundle of fibres

fibre

myofibril

filaments

Inside a muscle

This stylized illustration shows how a muscle is made of bundles of *fibres* ...

And each fibre is made of strands called *myofibrils* ...

And each myofibril is made of two kinds of interlocking *filaments*.

How muscles pull

Your muscles pull your bones into different positions by *contracting*. During contraction, the tiny interlocking filaments inside the muscle move closer together. When contraction is complete, the filaments are held in position by chemical links.

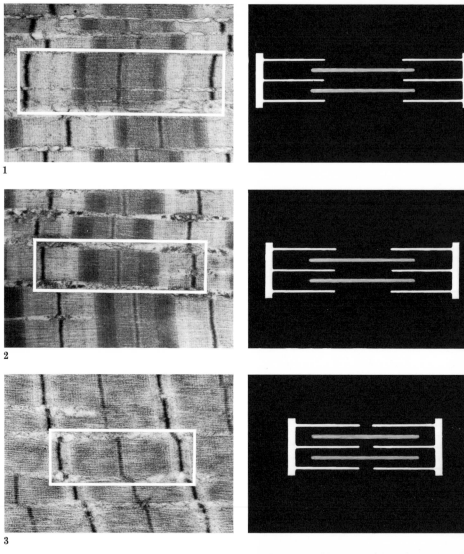

1

2

3

These photographs and diagrams show what happens to the filaments as a muscle contracts.

Compare the photographs with the diagrams. You can see the dark bands getting closer together as the filaments slide past each other.

4   Photographs 14 000 times lifesize.

What makes muscles pull? – Nerves

Your muscles pull when they receive *signals* from your brain telling them to do so. These signals are carried by *nerves* . . .

Inside a nerve

Nerves are made of *neurons* (nerve cells). Each neuron is specially shaped to receive, carry and pass on signals.

The branches of the *cell body* increase its surface area, so it can receive signals from many other neurons.

Special cells around the long *axon* of the neuron help it carry signals.

The flattened *end plates* enable the neuron to pass signals to the muscle.

This simplified drawing of neurons is not to scale – the axons should be very much longer.

You can see some photographs of neurons on page 48.

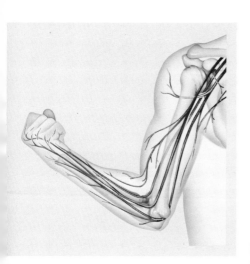

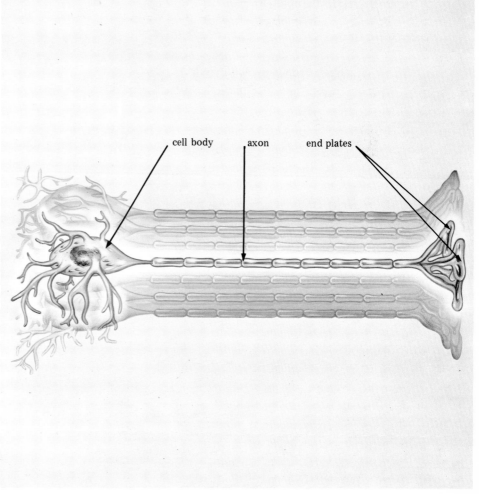

cell body   axon   end plates

Nerve signals

The cell body of a neuron receives signals from other neurons. When it is receiving enough signals, the neuron will 'fire' and send signals along its axon – either to a muscle, or to another neuron.

Nerve signals involve complex electrical and chemical changes in and around the neuron.

In the following diagrams, these complex changes are represented by different colours.

Sending signals to a muscle

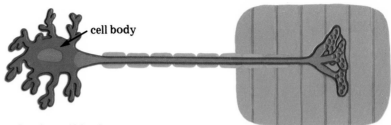

1 *In the cell body*
When enough signals are reaching the cell body, they cause electrical changes (red) inside it.
These electrical changes trigger off new signals.

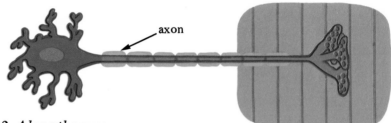

2 *Along the axon*
The signals (red spots) travel along the axon in a series of jumps. They are reinforced at each of the gaps between the surrounding cells.

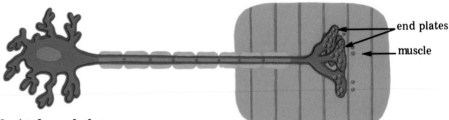

3 *At the end plates*
When the signals reach the end plates, a chemical (blue) is released. This chemical passes to the muscle.

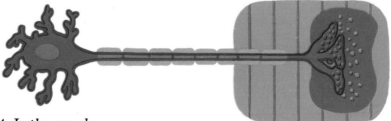

4 *In the muscle*
The chemical causes electrical changes (red) in the muscle. These electrical changes cause the muscle to contract.

Passing signals to other neurons

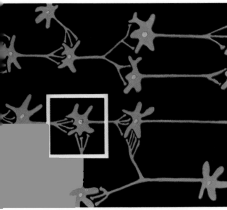

In your body, there are millions of neurons of many different kinds. These neurons are all linked together to form a complex signal-carrying network.

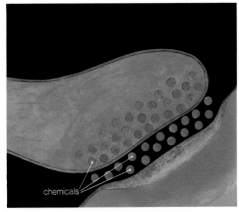

chemicals

3 When a signal arrives at a synapse, stored chemicals (blue) are released into the gap. These chemicals cross to a special surface on the next neuron.

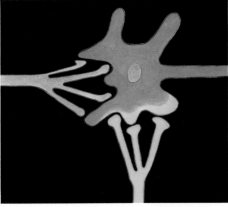

5 Other synapses store *inhibitory* chemicals. If enough of these chemicals reach the next neuron, they cause electrical changes (yellow) of a different kind.

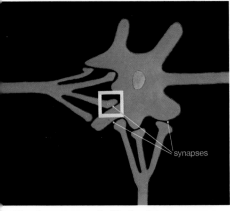

synapses

Each neuron is linked to many others, but it is not actually joined to them. There is always a small gap between one neuron and the next. The point where a signal crosses this gap is called a *synapse*.

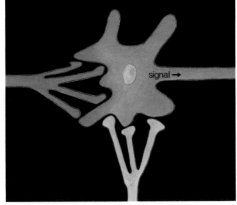

signal →

4 Some synapses store *excitatory* chemicals. When enough of these chemicals have reached the next neuron, they cause electrical changes (red), which trigger off new signals. This is how signals are passed from one neuron to another.

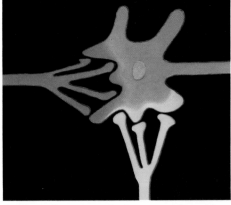

6 Signals can only be passed on if the second neuron receives more excitatory chemicals than inhibitory ones.

Synapses help to control the way signals are passed through the network of neurons in your body.

You can find out more about passing on signals in the next chapter.

The nervous system

Your nervous system is a network of millions of neurons. Most of these neurons are in your *brain* and *spinal cord* – your *central nervous system*.

Other neurons make up the nerves which carry signals to and from all parts of your body. Each nerve carries signals in one direction only – some nerves carry signals *to* the central nervous system, others carry signals *from* the central nervous system to other parts of your body.

The brain

Your brain is the coordinating and decision-making centre of your nervous system. It is made of millions of neurons, arranged in elaborate networks. There are so many complex interactions between these networks that scientists are only just beginning to understand how the brain works.

Different areas of the brain seem to be responsible for different things.

For example, the *cortex* – the outer layer of the brain – receives and processes signals and controls your actions.

You can find out more about the cortex in the next chapter.

The *cerebellum* controls your balance and posture.

The *hypothalamus* controls your responses to basic needs such as eating, drinking and sleeping.

You can find out more about the hypothalamus in Chapter 7.

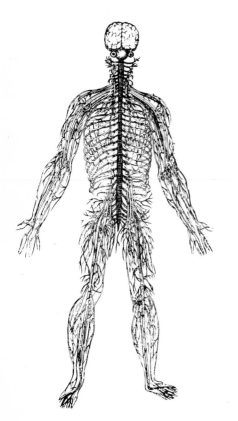

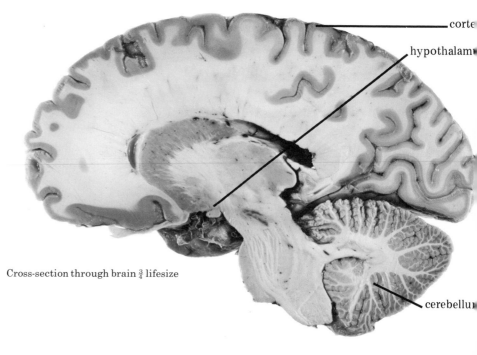

cortex

hypothalamus

cerebellum

Cross-section through brain ¾ lifesize

44

Chapter 5

Controlling your actions

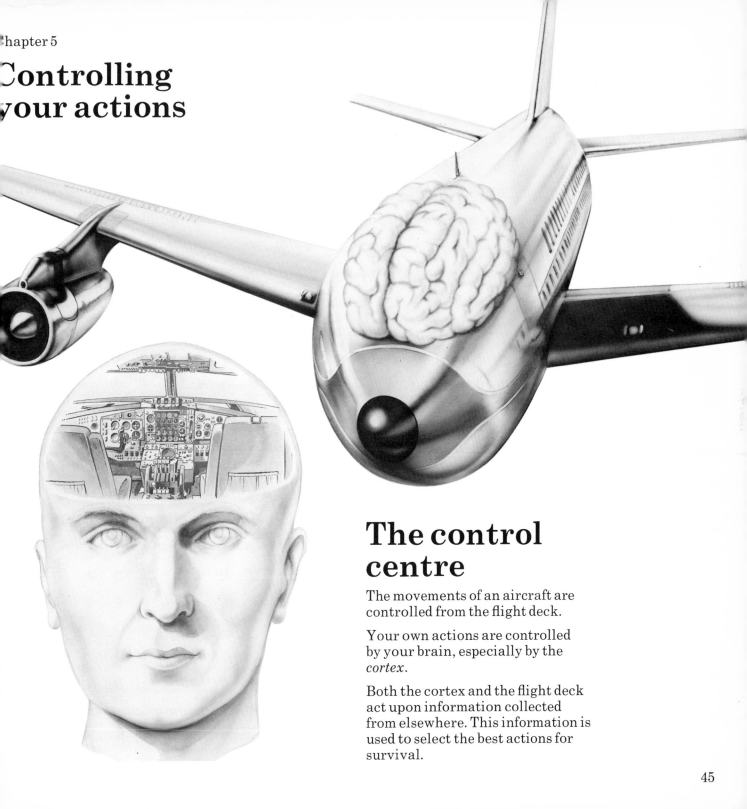

The control centre

The movements of an aircraft are controlled from the flight deck.

Your own actions are controlled by your brain, especially by the *cortex*.

Both the cortex and the flight deck act upon information collected from elsewhere. This information is used to select the best actions for survival.

Mapping the cortex

Different areas of your cortex are responsible for different things. The *sensory* areas collect information from your senses. The *association* areas process this information, and the *motor* areas send instructions to your muscles.

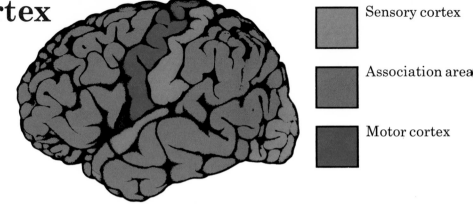

Sensory cortex

Association area

Motor cortex

The sensory cortex

Information about the outside world enters your nervous system through *receptors* in your sense organs and in your skin. These receptors convert all kinds of sensory information into nerve signals, which travel along nerves to your sensory cortex. Signals from the different senses are sent to different areas of the sensory cortex.

Within the 'touch and pain' area of the sensory cortex, different areas are responsible for signals from different parts of your body.

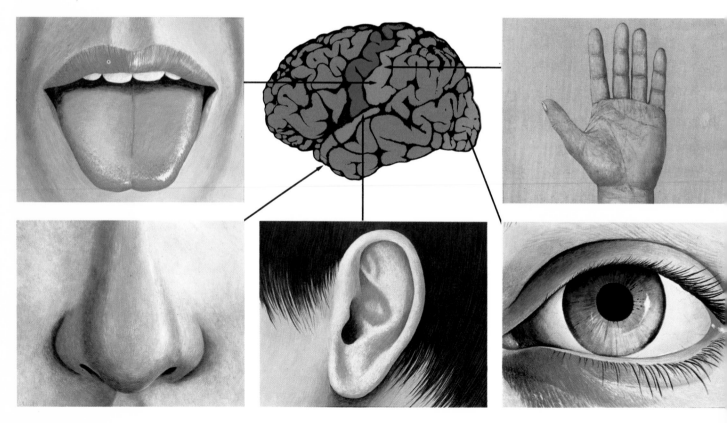

The association areas

Signals pass from the sensory areas to the association areas of the cortex. The association areas are responsible for understanding speech, and for ideas and decisions. They interpret and analyse the incoming signals, and decide how to respond to them.

The motor cortex

When you decide to carry out an action, signals are sent to the motor areas of the cortex. These areas then send out signals along nerves to the relevant muscles.

Signals to muscles in different parts of your body are sent from different areas of the motor cortex.

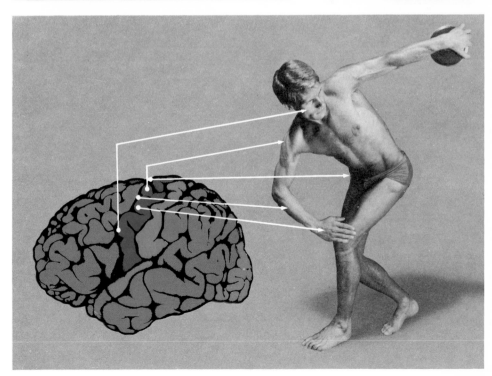

Inside the cortex

The cortex is like a very thin, deeply folded, sheet wrapped around the other parts of your brain. It is made up of more than 10 000 000 000 *neurons* (nerve cells).

1 Section through the cortex, $\frac{1}{3}$ lifesize.

2 A closer look at part of the cortex, 5 times lifesize.

3 Neurons of the cortex, 50 times lifesize. These have been stained by the Golgi method, which picks out only a tiny fraction of the neurons that are there.

4 A single neuron, 1000 times lifesize. This has been dissected out and photographed by a scanning electron microscope.

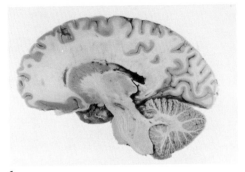
1

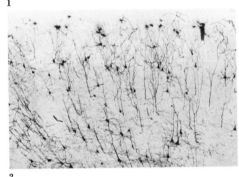

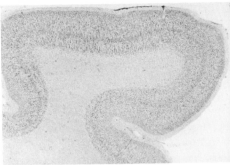
2

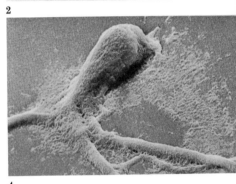
3

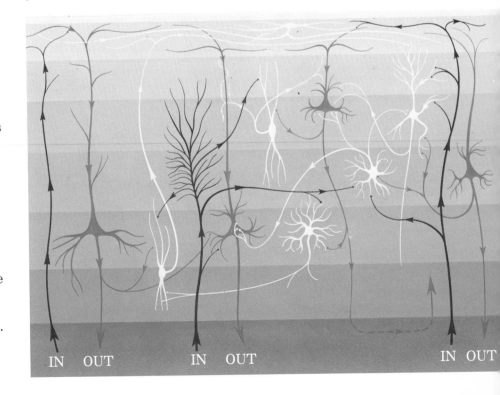
4

Neurons in the cortex

The neurons in your cortex have many different shapes and functions. Many of them are connected to neurons outside the cortex. Some of these deal with signals coming *in* from other parts of the body *(afferent signals)*.

Some deal with signals going *out* from the cortex *(efferent signals)*.

Other neurons deal only with signals within the cortex itself.

All these neurons are arranged in layers and columns, forming complex networks throughout the cortex. These networks provide countless millions of different pathways for the signals to follow.

IN OUT   IN OUT   IN OUT

48

Passing on signals

Each neuron gathers signals from other neurons in the network and passes them on, probably according to a simple 'voting' principle.

These diagrams show how such a voting system might be used to coordinate incoming signals.

Each neuron can receive signals only from the two neurons below it. When it is receiving signals from BOTH these neurons, it 'fires' and passes on the signals to the neuron above.

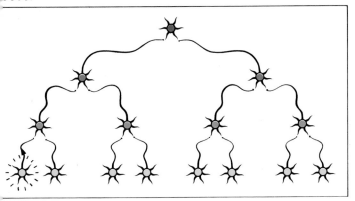

A yellow neuron 'fires' and sends signals to the green neuron above it.

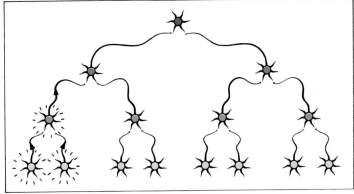

2 When the green neuron receives signals from both yellow neurons below it, it fires and passes on the signals to the blue neuron above it.

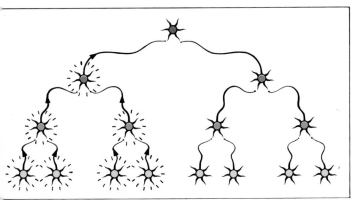

When all four yellow neurons on the left are firing, both green neurons fire, and the blue neuron passes on the signals.

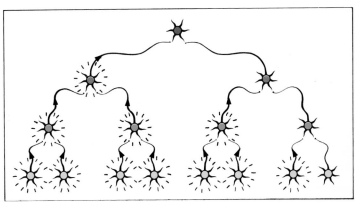

4 Which neuron must begin firing to enable the red neuron to fire?

49

The visual pathway

Information from your eyes passes to an area of the sensory cortex known as the *visual cortex*. The route it takes is known as the *visual pathway*.

Light from an object enters the eye and forms an image on the *retina*. Receptors in the retina convert this image into nerve signals, which travel along the *optic nerve* to the visual cortex.

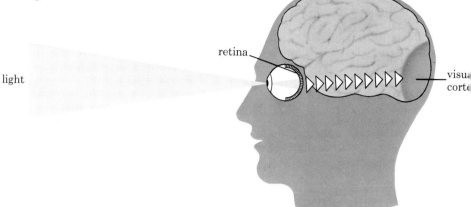

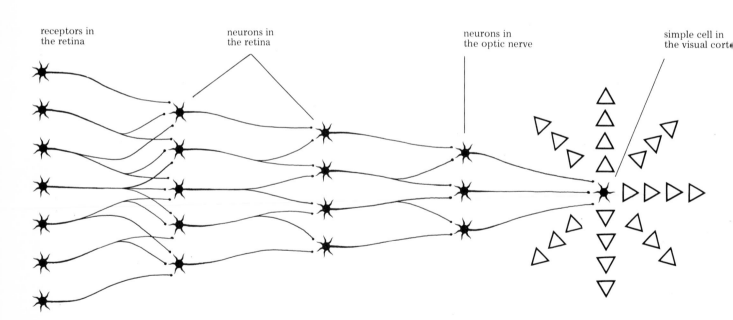

receptors in
the retina

neurons in
the retina

neurons in
the optic nerve

simple cell in
the visual cortex

The signals from the retina do not pass directly to the visual cortex. They are collected and passed on by the neurons which make up the visual pathway.

In the visual cortex, the signals are received by the *simple cells*, which pass them on to other neurons. As the signals pass on through the neuron networks of the cortex, they are analysed and interpreted. It is only when they have been analysed in this way that you actually perceive the object that is in front of your eyes.

You can find out more about perception in Chapter 11.

Basic needs

Your brain gathers information from inside as well as from outside your body. It uses this information to help satisfy your *basic needs*, such as hunger, thirst, curiosity and sex. So your actions are a response to information from inside as well as from outside your body – as the cartoon sequence shows.

Signals in
Input from other parts of the body to the cortex.

2 Signals in
Input from the senses to the cortex.

Ideas and decisions
Processing of information by the cortex.

4 Signals out
Instructions from the cortex to the muscles.

Ideas and decisions

What happens to all the information gathered by your cortex? You interpret it and decide how to respond to it. You store some of it as memories. You learn from it. You use it when you form ideas.

All these thought processes are carried out by large areas of your brain, including the association areas of your cortex.

You can find out more about memory and learning in Chapter 10.

Touch one of these spots.

Now read on . . .

Sensing, thinking, acting

Did you touch one of the spots?
Even simple actions like this
involve the three major areas of
your cortex.

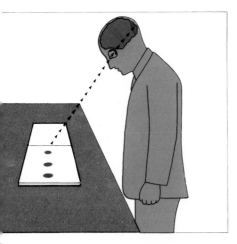

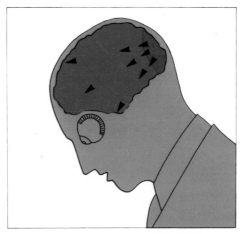

You see the three spots and the
words asking you to touch one of
them. Light from the spots and the
words enters your eyes and forms an
image on the retina at the back of
each eye.

Receptors in the retina convert this
image into nerve signals, which
travel along the visual pathway to
the visual cortex.

The visual cortex analyses the
information from your eyes and
passes it on to other parts of your
brain, including the association
areas of the cortex.

This is when you realize you have a
choice. You can choose to ignore
the instruction and the spots, or
you can choose to touch one of them
– in which case you have to decide
which one to touch.

When you have made your
decision, the motor areas of your
cortex send signals to the
appropriate muscles.

Controlling your actions

Many of your actions are carried out in response to information gathered by your brain. But this does not mean that you are like a puppet – passively waiting for information to reach you before you can do anything.

It is usually the other way round. You carry out most of your actions in order to gather information. This is most obvious in young children, who are always moving about looking for things to watch or explore.

You can find out more about the way children explore the world around them in Chapter 12.

Your actions are never as simple as a puppet's. Even a trivial action like turning the page of a book is made up of many small movements. As you make each one of these movements, your brain senses it, checks that it is right, and decides on the next one.

You become aware of these separate little movements when you try to learn new skills – like riding a bike, or playing a musical instrument – even feeding yourself.

You can find out more about learning skills in Chapter 10.

In almost everything you do, there is continuous feedback between your actions and the information gathered by your brain.

Chapter 6

Your life in the balance . . .

Inside your body, thousands of self-regulating processes are keeping you alive. These processes are helping to keep conditions stable.

The tendency for conditions inside your body to stay the same, even though conditions outside are constantly changing, is known as *homeostasis*.

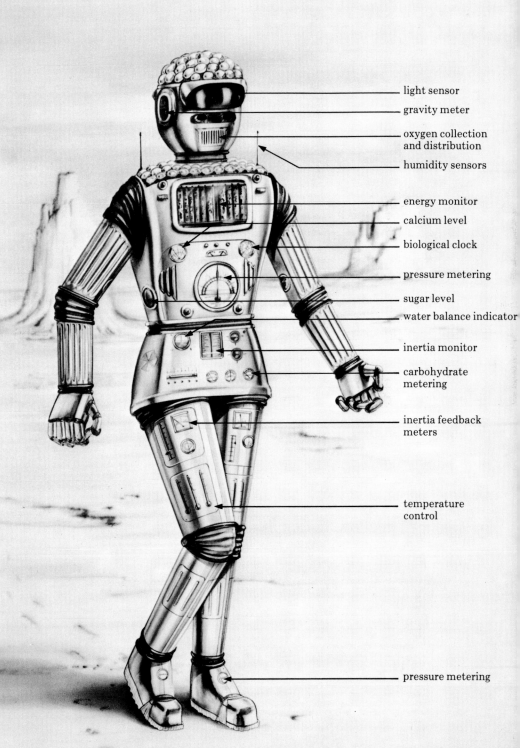

light sensor

gravity meter

oxygen collection and distribution

humidity sensors

energy monitor

calcium level

biological clock

pressure metering

sugar level

water balance indicator

inertia monitor

carbohydrate metering

inertia feedback meters

temperature control

pressure metering

55

Homeostasis

Homeostasis requires energy – indeed most of your energy is spent on keeping your body in a steady state.

Homeostasis involves a continuous exchange of energy and materials between you and your surroundings – but (in the short term) input always balances output, so everything appears to stay the same.

Homeostasis involves all the familiar body processes that keep you alive – breathing, eating, drinking and healing.

Do you need a drink?

Drinking is an example of homeostasis. You drink to replace the water you are continuously losing through sweating and other forms of excretion.

Drinking is part of a complex regulating system that controls the amount of water in your body.

You can find out more about water control in Chapter 9.

Responding to light

Light enters your eyes through the pupils. The size of the pupils affects the amount of light entering.

Have you ever noticed how your pupils respond to light? Look in a mirror. In dim light, your pupils enlarge. In bright light, they become small.

The way your pupils respond to light is an example of homeostasis. The change in size helps your eyes adjust to different light conditions.

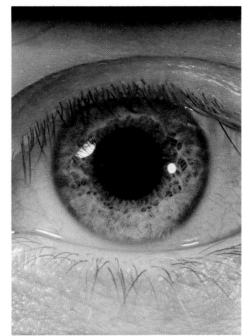

In dim light.

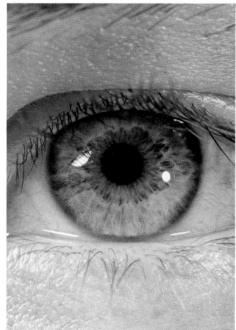

In bright light.

Resting is another aspect of homeostasis . . .

A broader view of homeostasis

Breathing, eating, drinking – these are all examples of *short-term homeostasis*. They keep you alive from one minute to the next.

Over longer periods, your survival is ensured by the repair mechanisms that heal your wounds and replace worn-out cells. These are examples of *long-term homeostasis*.

Learning and memory are less obvious examples of long-term homeostasis. Throughout your lifetime, they improve your chances of survival by enabling you to exploit your surroundings more effectively.

You can find out more about memory and learning in Chapter 10.

On a broader scale, sexual reproduction is a homeostatic process. Obviously it does not ensure your own survival. But it does ensure that, before death, you can be replaced by children. During your lifetime, the environment inevitably changes, and your children will have to face different conditions. Sexual reproduction produces children who are slightly different from their parents, and perhaps better able to survive these changed conditions. Sexual reproduction is therefore a homeostatic process that ensures the long-term survival of the human race.

Chapter 7
Hormones – messengers in the blood

Emergency!

How do you react?
What happens to your body?

Emergency!

When a large dog leaps over a gate towards you, barking savagely . . .
When you stumble on a river bank and fall towards the water . . .
When you are woken at night by a strange noise . . .
When you misjudge the speed of an approaching car . . .
Whenever you face an emergency, your body reacts in a definite way. It is always the same, whatever the emergency. These are some of the things that happen to your body – how do you think they help you survive?

Appearance Eyes open wider. Skin pales as blood is diverted to muscles.

Blood Clots faster. Carries more fuel to muscles.

Breathing Rate increases. Airways open wider.

Sweating Increases.

Action Muscles can work longer without tiring. Blood is diverted from gut to muscles. More fuel is made available for muscles.

The emergency reaction

All these things happen almost at once. They are part of the *emergency reaction*.

In the emergency reaction, messages must be sent to cells (*targets*) in all parts of your body. These messages cannot be carried by your nervous system, because nerves do not communicate with all your cells. Instead, they are carried in your blood, which travels all round your body and contacts every cell.

The messages are carried in your blood by a chemical called *adrenalin*. As adrenalin is carried round your body, it reaches almost all your cells. But they do not all respond. It is as if adrenalin carries a coded message which only certain cells (the targets) can decode. Adrenalin has many different target cells, including muscle cells, gut cells and skin cells.

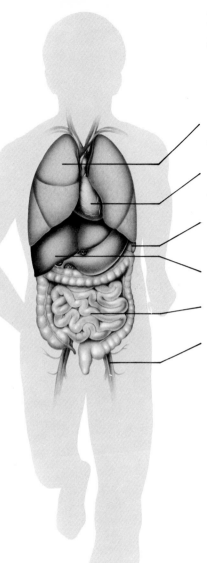

How does adrenalin help you survive an emergency?

Lungs *Wider airways let in more air. Faster breathing provides more oxygen for muscles.*

Heart *Faster, harder pumping pushes more blood (containing fuel and oxygen) to the muscles.*

Spleen *More red blood cells are released, so blood can carry more oxygen.*

Liver *Stored food is released to provide more fuel for muscles.*

Gut *Blood from gut is diverted to muscles, so they can work harder.*

Blood *If there is an accident, faster clotting reduces bleeding.*

Sweating *Increased sweating cools hot working muscles.*

Pale skin *Blood from skin is diverted to muscles, so they can work harder.*

During the emergency reaction, adrenalin carries messages to cells in all parts of your body. As a result, your muscles and nervous system can work harder, and you are better prepared to face the emergency.

Hormones

Adrenalin is only one of many chemical messengers in your body. These chemical messengers are known as *hormones*.

Most hormones work in the same way as adrenalin. You cannot consciously control them. They do not follow specific routes to their targets, but reach them by chance as they travel round your body in the bloodstream.

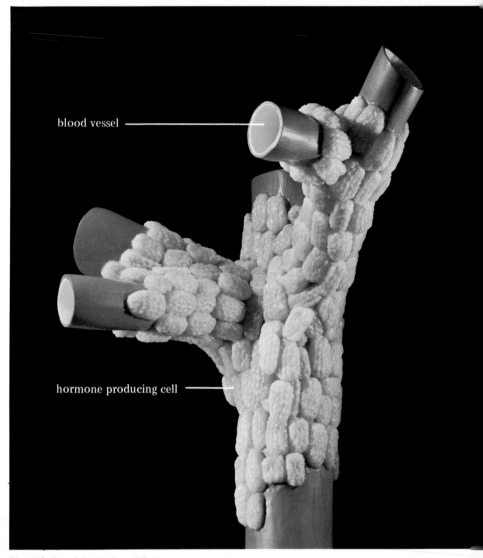

blood vessel

hormone producing cell

Simplified model, 500 times lifesize

Where do hormones come from?

Hormones are manufactured by special cells grouped together to form *glands*. Many of these glands have familiar names (pancreas, ovaries, thyroid), and most of them produce more than one hormone.

The hormone-producing cells are attached to thin-walled blood vessels, and hormones are released directly into the bloodstream.

What do hormones control?

Hormones are not released continuously. They are released in short bursts which vary in quantity and frequency. And different cells respond at different speeds. This means that hormones can be used to control a wide variety of processes, ranging from the high-speed emergency reaction to the slow process of growth.

Because hormones *can* govern slow processes, they are used to control many important body processes . . .

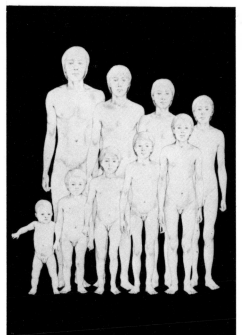
Growth

Water balance

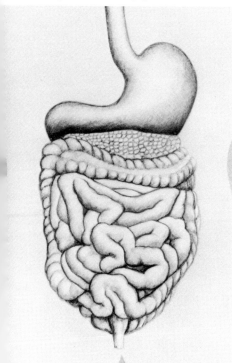
Digestion

Milk production

Reproduction and the development of sexual features

63

The hormone-producing glands

Most glands produce several hormones. And some of these hormones control many different processes. Here are the most important hormone-producing glands, together with some of the hormones they produce, and the main processes they control.

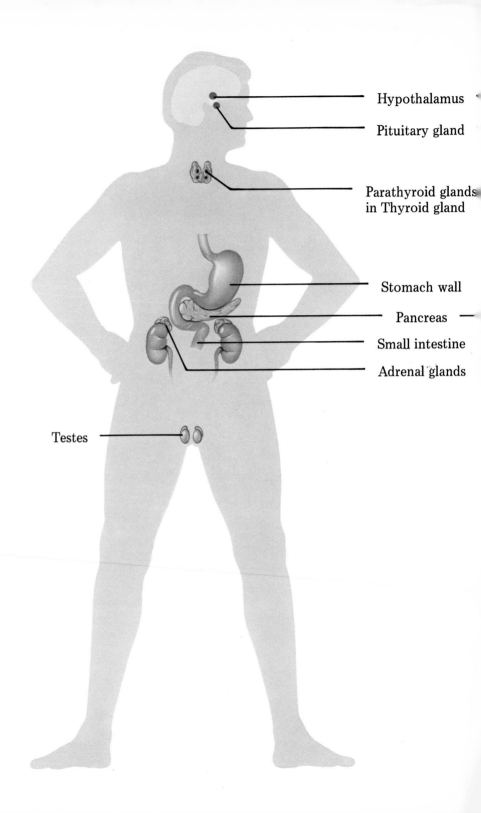

Hypothalamus

Pituitary gland

Parathyroid glands in Thyroid gland

Stomach wall

Pancreas

Small intestine

Adrenal glands

Testes

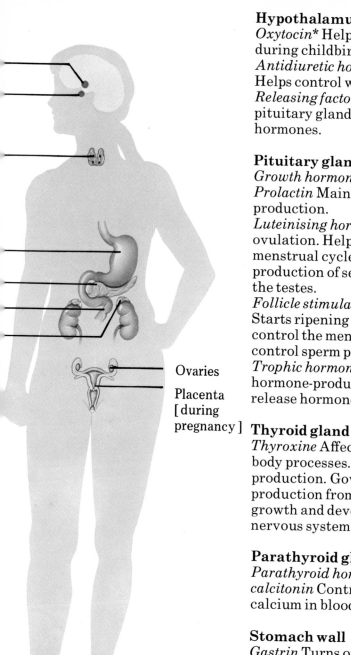

Hypothalamus

Oxytocin* Helps uterus contract during childbirth.

Antidiuretic hormone* (ADH) Helps control water balance.

Releasing factors Stimulate the pituitary gland to release hormones.

Pituitary gland

Growth hormone

Prolactin Maintains milk production.

Luteinising hormone (LH) Triggers ovulation. Helps control the menstrual cycle. Controls production of sex hormones from the testes.

Follicle stimulating hormone (FSH) Starts ripening of the ova. Helps control the menstrual cycle. Helps control sperm production.

Trophic hormones Stimulate other hormone-producing glands to release hormones.

Thyroid gland

Thyroxine Affects rate of general body processes. Controls heat production. Governs energy production from food. Controls the growth and development of the nervous system.

Parathyroid glands

Parathyroid hormone and *calcitonin* Control amount of calcium in blood and bones.

Stomach wall

Gastrin Turns on acid production by stomach.

Small intestine

Enterogasterone Turns off acid production by stomach.

Secretin and *CCK-PZ* Trigger release of digestive enzymes from pancreas.

Pancreas

Insulin Helps control amount of sugars in blood.

Adrenal glands

Adrenalin Controls emergency reaction.

Cortisol Helps body cope with stress. Helps control water balance.

Aldosterone Helps control water balance.

Androgens Help control development of male features at puberty. Help control beard growth.

Testes

Testosterone Helps control sperm growth and development. Helps control development of male features at puberty. Helps control beard growth.

Ovaries

Progesterone Helps control normal progress of pregnancy. Interacts with FSH, LH and oestrogen to control the menstrual cycle.

Oestrogen Controls development of female features at puberty. Interacts with FSH, LH and progesterone to control the menstrual cycle.

Placenta (during pregnancy)

HCG and *HCS* Help control normal progress of pregnancy.

Oestrogen and *progesterone* Start milk production.

*Stored in pituitary gland.

Ovaries

Placenta [during pregnancy]

The glands in control

Hormones cannot control body processes efficiently unless they are maintained at the right level in the blood. Hormone levels are controlled by the *hypothalamus* and the *pituitary gland*, working together. These are the real controllers of the hormone system. By regulating the amount of each hormone present in the blood, they indirectly govern all the body processes governed by hormones.

Feedback

The hypothalamus and the pituitary gland control hormone levels in the blood by a balancing mechanism known as *feedback*. They use feedback in many different ways.

A simple feedback loop
One of the simplest feedback systems is the *negative feedback loop*. This is the system that is used to maintain thyroid stimulating hormone (TSH) at the required level in your blood. Your hypothalamus detects when the amount of TSH is approaching the required level, and commands your pituitary gland to decrease the amount of TSH released.

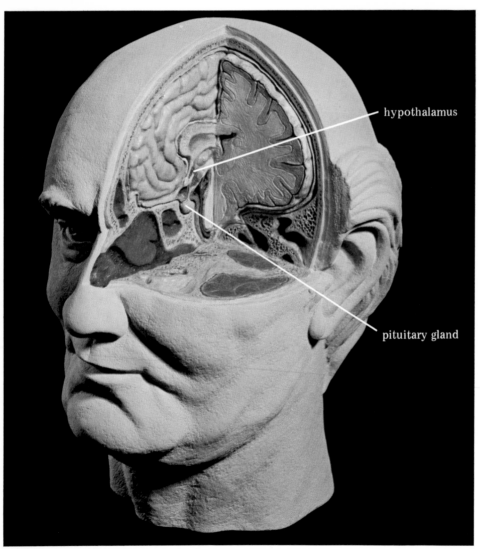

hypothalamus

pituitary gland

The pituitary gland is closely linked to the hypothalamus, which is part of the brain itself.

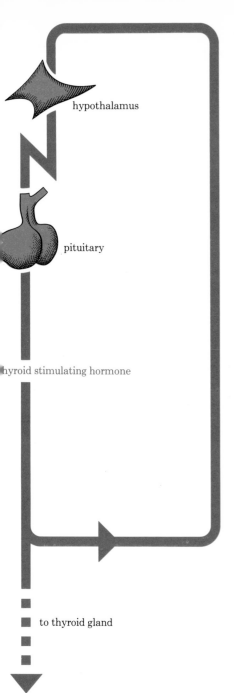

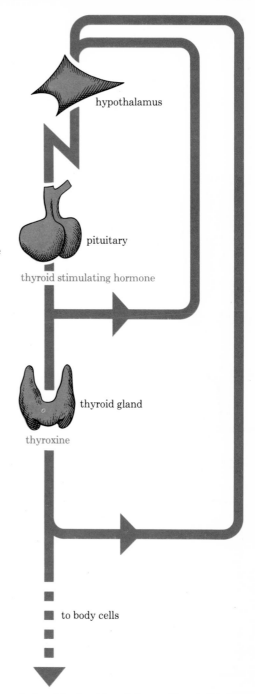

A double feedback loop

Many of the feedback systems involved in hormone control include two or more loops linked together in some way. For example, the feedback system that controls the level of the hormone thyroxine is a *double loop* – a loop within a loop.

The inner loop is the simple negative feedback loop that controls the amount of TSH in the blood. TSH stimulates the thyroid gland to release thyroxine. But an increase in the amount of thyroxine in the blood causes the pituitary gland to release less TSH – this is the outer loop. Both loops indirectly control the level of thyroxine by regulating the amount of TSH present in the blood.

A simple feedback loop. As the amount of TSH in the blood increases, less is released.

A double feedback loop. Part of each loop (indirectly) turns off the thyroid gland.

67

Feedback systems
Nearly all the body processes
controlled by hormones involve
very complex feedback systems.
Often several hormones are
involved, some working together
(this is known as *synergism*) and
some working against each other
(this is known as *antagonism*).

synergism

antagonism

Your hormone system is an
incredibly complex, delicately
balanced, network, and scientists
are only just beginning to
understand how it works.

Chapter 8

More about sex hormones

How do you tell the sex of a newborn baby?

A girl has a vagina. A boy has a penis. And, although they are not visible, there are other sex organs inside the baby's body.

The way these sex organs develop before birth is controlled by *sex hormones*.

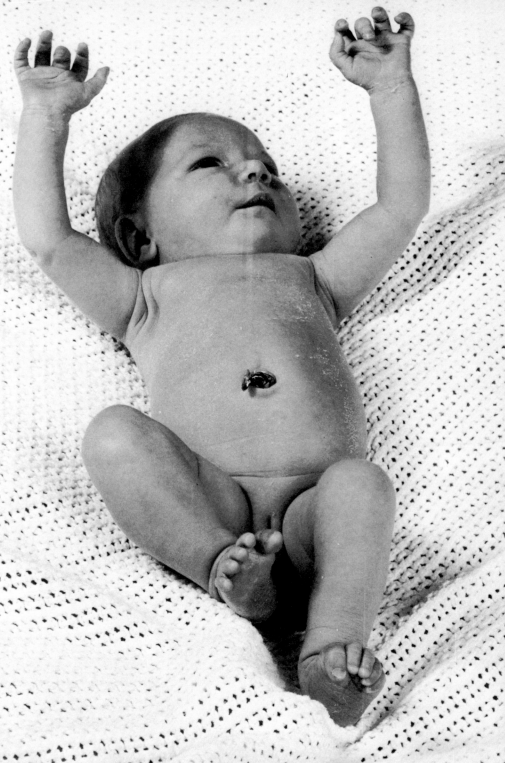

Sex hormones before birth

A baby's sex is decided at fertilization by the two *sex chromosomes (described on page 34)* it receives from its parents.

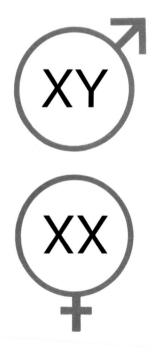

A baby boy receives one X and one Y sex chromosome. A baby girl receives two X sex chromosomes.

1 For the first six weeks after fertilization, both sexes develop in the same way.

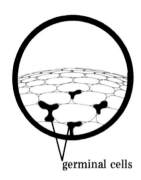

germinal cells

2 A group of special *germinal cells* is formed. If the baby is a boy, these cells will one day produce *sperm cells*. If it is a girl, they will develop into *ova*.

3 The germinal cells move towards two cell ridges – the *genital ridges* – near the baby's developing kidneys.

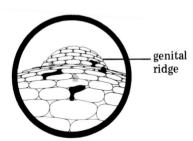

genital ridge

4 At this stage, these ridges can develop into EITHER male *testes* OR female *ovaries*.

5 If the baby received an X and a Y sex chromosome at fertilization, the genital ridges now start to develop into testes.

6 The germinal cells move in amongst the cells of the genital ridges to become part of the testes. Some of the other cells in the developing testes now begin to produce *male sex hormones*.

7 At this stage, the cells near the developing testes can develop into EITHER a male *penis* and *sperm ducts* OR a female *vagina, uterus* and *oviducts*.

8 They wait for a critical length of time . . .

9 If these cells receive male sex hormones from the developing testes *before* the end of this critical period,

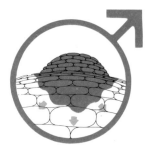

10 they develop into a penis and sperm ducts.

1 Six or seven months later,

2 a baby boy is born.

13 BUT, if the baby received two X sex chromosomes at fertilization, the genital ridges start to develop into ovaries.

14 The germinal cells move in amongst the cells of the genital ridges to become part of the ovaries.

15 At this stage, the cells near the developing ovaries can develop into EITHER a male penis and sperm ducts OR a female vagina, uterus and oviducts.

16 They wait for a critical length of time . . .

17 As there are no developing testes, there are no male sex hormones being produced. Because they do not receive male sex hormones during the critical period,

18 the cells near the developing ovaries automatically begin to develop into a female vagina, uterus and oviducts.

19 Six or seven months later,

20 a baby girl is born.

Sex hormones during childhood

Although you continue to produce sex hormones after you are born, they have little effect on your sex organs until you reach puberty.

No one really knows what sex hormones do during childhood.

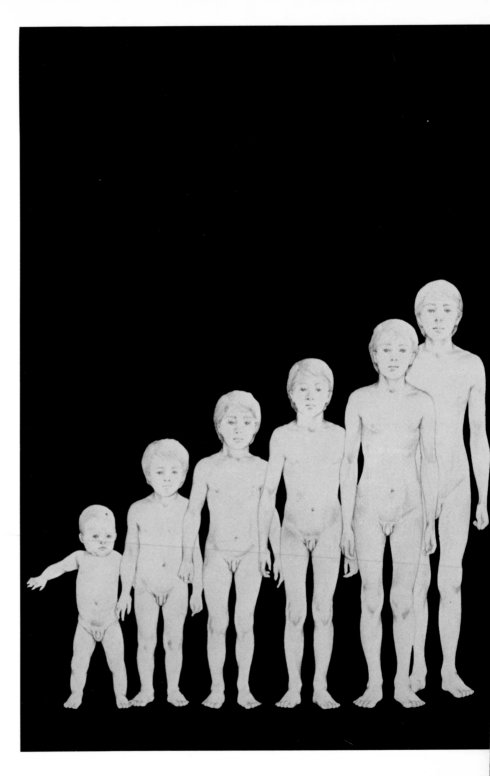

Sex hormones and puberty

When you reach puberty . . . Your brain turns you on.

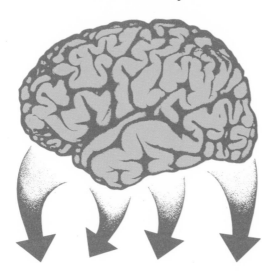

Puberty begins when your hypothalamus switches on your ovaries and testes to prepare your body for reproduction.

How does it do this?
By telling your pituitary gland to increase its production of the hormones FSH and LH, which stir your ovaries or testes into action.

What do your ovaries and testes do?
From puberty onwards . . .
1 Ovaries produce mature ova. Testes produce sperm cells.
2 Ovaries and testes both release large amounts of sex hormones, which control the development of sexual features.

1 Hormones, ova and sperm cells
The production of ova and sperm cells is controlled by the hormones FSH and LH from the pituitary gland.

In males, FSH and LH are released more or less continuously, resulting in a continuous production of sperm cells.

In females, FSH and LH are released at regular intervals as part of the *menstrual cycle* ('monthly periods'). Each month, they help to control the growth and release of one ovum from an ovary.

You can find out more about the menstrual cycle later in the chapter.

2 Hormones and sexual features
Ovaries produce *female sex hormones,* including *oestrogen* and *progesterone*. These hormones control the development of female sexual features, such as breasts and rounded hips and thighs.

Testes produce *male sex hormones,* including testosterone. These hormones control the development of male sexual features, such as body hair and large, heavy muscles.

But males also produce small amounts of female sex hormones, and females produce small amounts of male sex hormones.

The way you look after puberty depends on a complex interaction between all your hormones and their target cells.

testes

ovaries

Before and after puberty

What features have changed?
In what way?
What are the changes for?

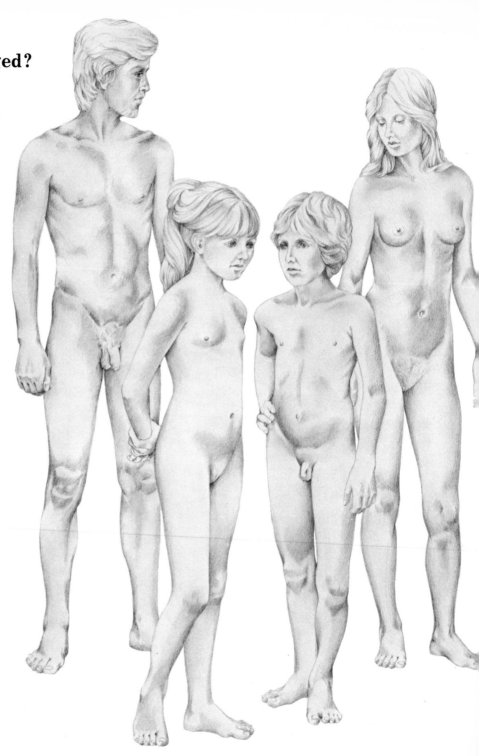

Changes you can see

In both sexes
● size increases dramatically
● pubic hair and body hair start to grow

In the male
● beard starts to grow
● jaw and neck change shape
● shoulders and chest enlarge
● muscles develop further
● penis, testes and scrotum enlarge

In the female
● breasts develop
● hips widen
● hips and thighs become rounded

Internal changes

In both sexes
● sexual drive begins to develop (This is triggered by hormones, then kept going by the brain.)

In the male
● the sperm ducts grow and develop
● the larynx (Adam's apple) enlarges, and the voice breaks
● the prostate gland enlarges and starts to produce seminal fluid

In the female
● the vagina, uterus and oviducts grow and develop
● the menstrual cycle ('monthly periods') begins

What are the changes for?

The main purpose of all the body changes at puberty is to get a mature ovum and sperm cell together, and to prepare the female body for a baby.

Many sexual features, such as the female breasts, are important *sexual signals*. They allow the sexes to recognize and attract each other.

The menstrual cycle

A beautifully balanced system

The female *menstrual cycle* ('monthly periods') is controlled by a beautifully balanced feedback system involving *four* different hormones.

The hormones involved are *FSH* and *LH* from the pituitary gland, and *oestrogen* and *progesterone* from the ovaries.

These four hormones work together to ensure that, each month, an ovum is released at the right time and in the right place for possible fertilization by a sperm cell.

In these diagrams, the hormones are represented by coloured dots – a different colour for each hormone.

The greater the number of dots, the greater the amount of hormone in the woman's blood.

Key

 hormones

⬤ where hormone comes from

✦ where hormone acts.

1 The menstrual cycle begins when the pituitary gland releases FSH (purple dots) into the bloodstream.

2 The increasing amount of FSH in the blood stimulates the growth of an ovum in one of the ovaries.

Oestrogen is released

3 As the ovum grows, the ovary begins to release oestrogen (blue dots) into the bloodstream.

4 The rising level of oestrogen in the blood causes two things to happen:
- less FSH is released – so no more ova will develop
- the lining of the uterus thickens to be ready to receive the fertilized ovum.

The ovum continues to ripen, and he increasing amount of oestrogen ventually triggers the release of H (pink dots) by the pituitary land.

7 The 'yellow body' releases progesterone (green dots) into the bloodstream. Progesterone prevents the pituitary gland from releasing either LH or FSH.

1 The falling level of progesterone allows the pituitary gland to start releasing FSH again.

Even while menstruation is continuing, the cycle begins again.

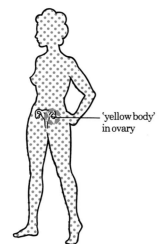

'yellow body' in ovary

How 'the pill' works

Most *contraceptive pills* contain some form of *oestrogen* and *progesterone*. Taking the pill regularly produces an artificially high level of these hormones in the blood and prevents the pituitary gland from releasing FSH or LH. This means that a ripe ovum is not released to be fertilized.

The oestrogen in the pill causes the lining of the uterus wall to thicken. When the pill is stopped for a few days, this lining breaks up and menstruation occurs.

The sudden flow of LH in the lood causes the release of the ripe vum from the ovary into the earby oviduct. This is known as *vulation*.

n the ovary, the small sac which eld the ovum begins to develop nto a *'yellow body'* (corpus luteum).

8 If the ripe ovum in the oviduct is *not fertilized* within a few days, the 'yellow body' stops producing progesterone and breaks up.

As the level of progesterone falls, the thickened lining of the uterus breaks up and *menstruation* (bleeding or 'a period') occurs.

Hormones and pregnancy

Pregnancy begins when a sperm cell fertilizes an ovum inside the mother. To allow the baby to develop in the lining of the uterus, the mother's menstrual cycle must be stopped . . .

After two weeks
The *placenta (described on page 22)* has begun to develop and is producing the hormones *HCG* and *HCS*. These hormones seem to be responsible for preventing the break-up of the 'yellow body' (corpus luteum) in the ovary.

The 'yellow body' continues to produce the hormone progesterone, which stops the menstrual cycle by:
● preventing the release of any more ova from the ovaries
● preventing the shedding of the uterus lining.

The placenta itself begins to produce the hormones progesterone and oestrogen.

After three months
The normal progress of pregnancy is completely under the control of the placenta, which is producing large amounts of progesterone and oestrogen. These hormones work together to control the growth of the uterus and prepare the breasts for milk production.

Progesterone also somehow prevents the uterus muscles from contracting and dislodging the developing baby.

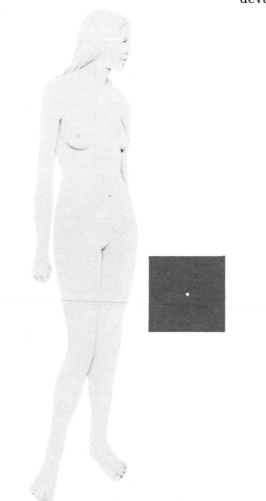

After eight months

The level of oestrogen in the mother's blood has risen so high that it is beginning to counteract the effect of progesterone on the uterus muscles. Oestrogen begins to prepare the uterus muscles for labour (described on page 20).

Hormones are responsible for almost all the changes in the mother during pregnancy.

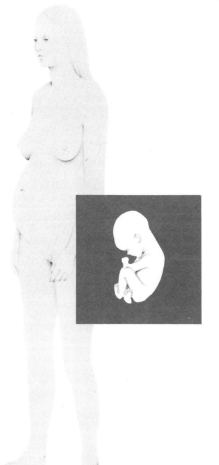

Hormones and birth

Childbirth (described on page 20) is triggered by the baby and the placenta, NOT by the mother.

When the baby is ready to be born . . .

Its hypothalamus seems to trigger a chain of events which results in the release of a hormone by the placenta. This hormone crosses into the mother's blood and is somehow responsible for the start of labour.

During labour

The baby's movements stretch the muscles of the uterus, sending nerve signals to the mother's brain. These signals trigger the immediate release of the hormone *oxytocin* from her pituitary gland. Oxytocin causes contractions of the uterus, leading to more movements of the baby – and the cycle repeats itself – over and over again – gradually becoming faster and more intense . . . until the baby is born.

YOU ARE 70% WATER

YOU ARE LOSING WATER ALL THE TIME, AND THE AMOUNT OF WATER IN YOUR BODY NEEDS TO BE CAREFULLY MAINTAINED.

HOW IS THIS DONE?

SO YOU USUALLY DRINK BEFORE YOU **NEED** TO - BEFORE YOU FEEL THIRSTY.

BUT WHAT HAPPENS WHEN YOU DO FEEL THIRSTY?

YOU HAVE A SYSTEM WHICH CONTINUALL MONITORS THE AMO OF WATER IN YOUR BODY.

THIS SYSTEM MAIN DEPENDS ON SPECIALI NERVE CELLS CALLE OSMORECEPTORS i THE HYPOTHALAMU

IF YOUR BODY LOSES ENOUGH WATER TO REDUCE THE VOLUME OF YOUR BLOOD, TWO OTHER HORMONES (ANGIOTENSIN AND ALDOSTERONE) ARE RELEASED. THESE WORK WITH A.D.H TO SAVE WATER BY ACTING ON THE KIDNEY AND THE BRAIN.

THE HYPOTHALAMUS SENDS A SIGNAL TO THE PITUITARY GLAND WHICH RELEASES THE HORMONE A.D.H. INTO THE BLOODSTREAM. THE BLOOD CARRIES THE HORMONE TO THE KIDNEY WHICH RESPONDS BY REDUCING THE AMOUNT OF WATER IN THE URINE.

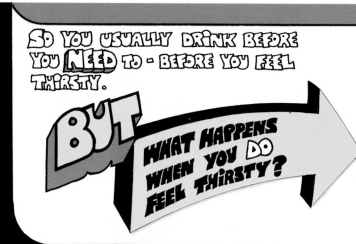

HOW IS THIS DONE?

IF YOU STOPPED SWEATING - YOU'D OVERHE

IF YOU STOPPED BREATHING - YOU'D DIE !

BUT EXCRETION CAN BE REDUCE

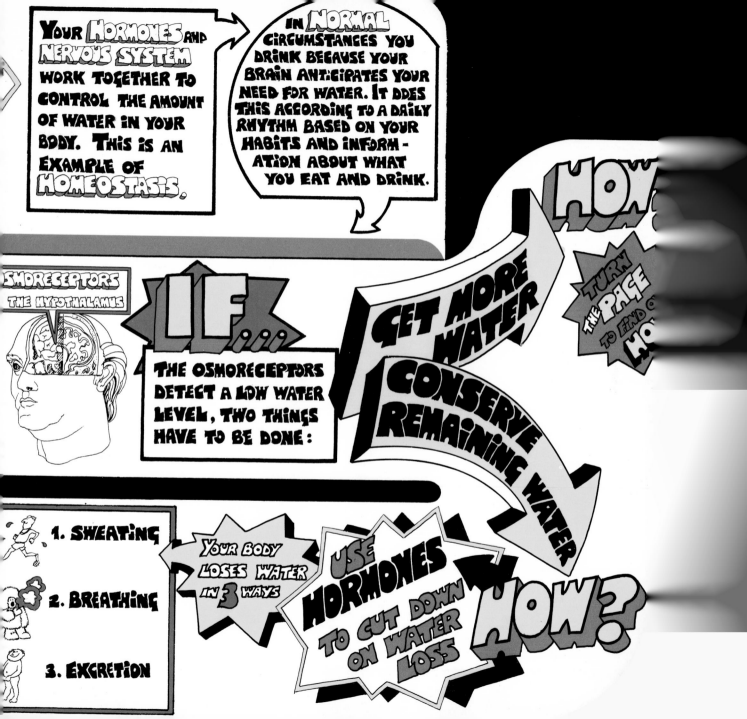

chapter 10

Experience of a lifetime

We don't just learn in classrooms . . .

We are learning all the time. We learn by organizing the information gathered by our senses.

In this way, we develop an understanding of the world. As our understanding changes, we change the way we do things – this is also learning.

We learn skills . . .

Learning to do something new – even something very simple like flipping a coin or hitting a nail with a hammer – involves your brain. It involves making new links between information received from your senses and the instructions sent to your muscles. These new links change the way you do things – do you handle a spoon in the same way now as you did when you first tried to feed yourself?

During our lives, we all acquire many skills. Every time we learn a new skill, we make countless new links between our senses and our muscles.

Learning and experience

What would you do in this situation?

What sort of things have you learnt that would influence your decision?

When did you learn these things? How?

We usually draw on a wide range of experiences when we learn. And it may be a long time before what we learn changes the way we do things

We learn by organizing our experiences . . .

Sorting the figures

You learn about the world from your experiences. In your brain, you *organize* the information you get from these experiences. Much of this information comes from real objects which you can experience directly through your senses.

One way that you organize such information is to sort it into groups based on similarities and differences.

Look at the figures and try to find similarities and differences between them. How would you sort them into groups?

Because they are all people, you could put all the figures together in one group.

Or you could sort them into two groups.

Or you could group them like this . . .

or this . . .

When we organize the information we get from our experiences, we are learning – though we may not always realize it.

or this . . .

or . . .

85

Find the krugsips

One way of organizing things in your brain is to sort them into groups. Here is another way of organizing things – by finding out how they are related to each other.

Look at these rows of shapes. There is a krugsip in every row. The krugsips in the first six rows are red. Can you find the krugsips in the last two rows?

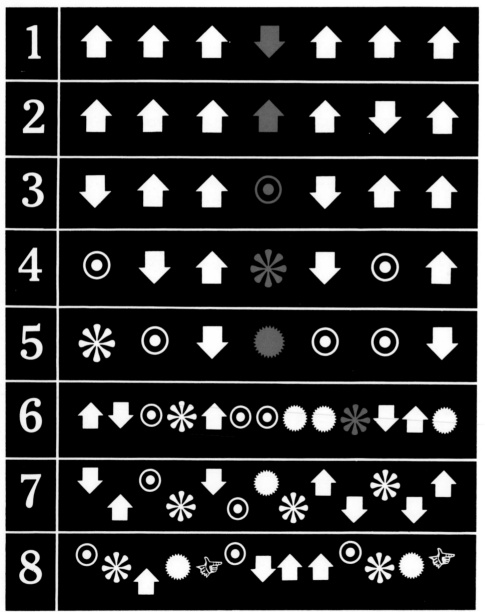

Krugsips are fourth from the right. Shape doesn't matter. It is the position which is important. Krugsip is a name for a *relationship* between things.

'Krugsip' is not a real word, but there are many real words which describe similar relationships – words like centre, top, outside and beginning. Some words, like 'truth' and 'hatred', describe very complicated relationships...

We often organize our experiences according to the relationships between them. When we do this, we are learning – though we may not always realize it.

We develop an understanding of the world

In your brain, you organize your experiences and store them in an orderly way. All these stored experiences make up your *understanding of the world*.

This understanding influences everything you do. Indeed you often do things just to test and improve your understanding.

When you learn, your understanding changes as you fit new experiences into it. And, when your understanding changes, you change the way you do things.

How do you learn?

You learn by doing things

There are many things that you cannot learn properly unless you do them yourself.

You learn by being told – with words

Words help you to learn fears, hopes and attitudes as well as detailed facts and instructions. They enable you to learn from people who are far away, or who are no longer alive. This means that your understanding of the world can include the experiences of previous generations.

You learn by being told – without words

You learn by observing

When we store our experiences as memories . . .

Try to remember your last meal . . .
Was the food hot or cold?
What did it smell like?
What did it look like?
Was it sweet or savoury?
Was it crunchy?

We store experiences from all our senses as memories.

We select what has meaning for us . . .

Why are some things easier to remember than others?

Look at the first group of words (below) for a while. Then look away and try writing the words down – or try repeating them to someone. Do the same with the other groups of words.

1
Drive and with yet should he meekly you rather timidly been again said as it one with glass began she ink to more said real.

2
Never imagine yourself not to be otherwise than what it might appear to others that what you were or might have been was not otherwise.

3
In order to find out what a book is all about, it is a good idea to read the chapter titles on the contents page.

4
Colourless sensible needles walk furiously between the tails which read enormously under a handkerchief while sandwiches think between laughing and sowing pianos in the mustard.

How much of each group of words do you remember? Group 3 is the easiest to remember, because it is the easiest to understand.

Do all the other groups of words make sense? It is difficult to remember something that does not make sense.

Look at these objects for half a
minute. Then look away and see
how many you remember.

How many did you remember?
You probably found it easy to
remember the objects that are
familiar to you. You may have
sorted the objects into groups. Or
you may have tried linking them
together, perhaps as a story.

One way or another, as you look
at the objects, you make them part
of your experiences. They then
have meaning for you – and are
easier to remember.

We recall what we have learnt . . .

Searching your memory

When you remember something, you are *recalling* it from your memory store.

You can recall the answers to some questions immediately. Other questions take longer to answer. This is because you have to search in your memory store before you can recall the answer.

How many windows are there in your bedroom?

One

And how many windows are there in your house?

. . . er . . . just a minute . . .

Helping you recall

When you search for something in your memory store, it is easier to recall if you have some hint or cue . . .

Do you remember Richard Brown?

No

Yes you do, he was at your party.

I can't think who you mean.

You must – he broke a window.

Oh, THAT Richard Brown. He was always clumsy. Do you know, he once . . .

Helping yourself recall

You can give yourself cues for recall. When you recall information, you can make use of the way the information is organized in your memory store. If you have organized the information as a sequence, you can use this sequence as a cue for recall.

My house
1...the Taylors
2...the Jacksons
3...don't know
4...the Parkers

Who lives in the house four doors down from you?

...er...just a minute...

...the Parkers

How would you recall the answers to these questions?...
What is a krugsip?
What were you doing this time last week?
Who lives four doors down from you?
How many days are there in July?

We all use our own methods, even for recalling the same information.

Solving problems

Even if you do not know the answer to a question, you may be able to work it out from the information in your memory store.

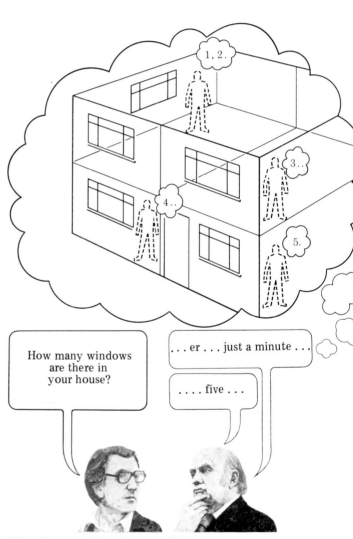

1, 2.

3..

4..

5.

How many windows are there in your house?

...er...just a minute...

....five...

We often combine items of information from our memory stores to produce 'new' information.

We choose actions which suit our needs

Learning makes us adaptable

People have learnt to live in different ways in different places...

Making the effort to learn

Learning needs effort. This is because you try to link what you are learning with what you have learnt before.

The effort seems less when you are learning something useful or interesting.

We can learn many ways of doing things and can choose the actions which best suit our needs.

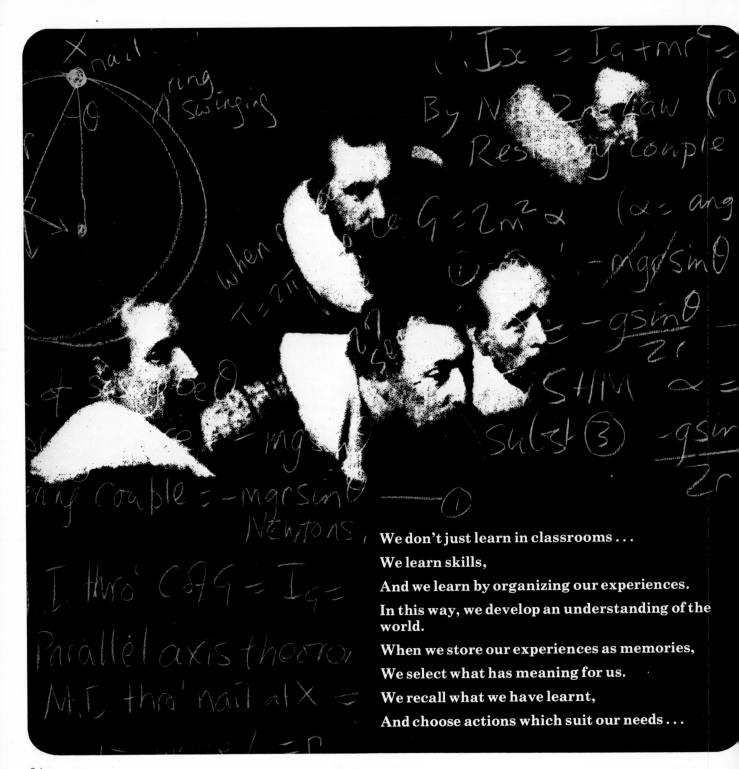

We don't just learn in classrooms . . .

We learn skills,

And we learn by organizing our experiences.

In this way, we develop an understanding of the world.

When we store our experiences as memories,

We select what has meaning for us.

We recall what we have learnt,

And choose actions which suit our needs . . .

Chapter 11

Perception understanding the world about us

How do you avoid walking into people in the street?
How do you recognize a fork?
How do you know how to make a cup of tea?

These are things that you do almost every day. They involve complex decisions based on what you see, hear, feel and know already.

Seeing, hearing, feeling and knowing

Knowing how large it is

Have you ever wondered how you know whether an object is small and near to you, or large and far away? The images formed in your eyes could well be the same size . . .

Fix an image (or rather an after-image) in your eyes by looking at a bright object for a moment or two (thirty seconds if it is not too bright). When you close your eyes, you will see an after-image of the object, imprinted in your eyes.

Now look at a distant flat surface, such as the wall at the end of the room. Then look at a flat surface nearer to you, such as the end of a cupboard. You will notice that the after-image appears first larger, then smaller. But the after-image is imprinted in your eyes. So it cannot change size.

Signals go from the after-image in your eyes to your brain. So it must be your brain that judges the size of the after-image according to its apparent distance from you.

Clues about distance often come from the background . . .
Are these three cubes all the same size?

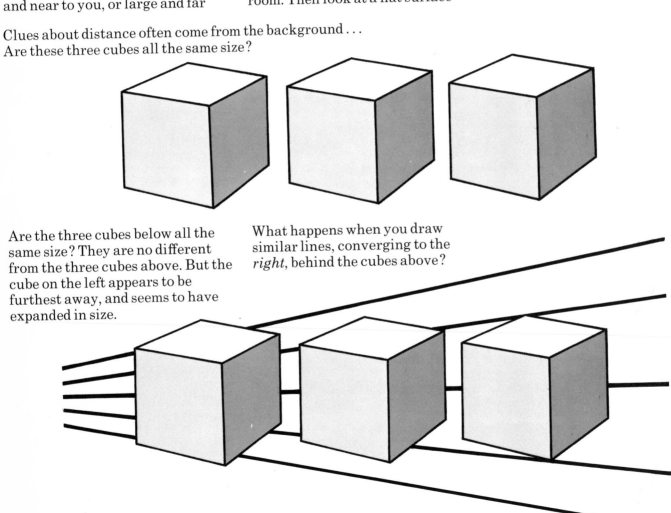

Are the three cubes below all the same size? They are no different from the three cubes above. But the cube on the left appears to be furthest away, and seems to have expanded in size.

What happens when you draw similar lines, converging to the *right*, behind the cubes above?

Seeing with two eyes

Have you ever noticed that what you see with one eye is slightly different from what you see with the other eye?

Look at a solid object – any solid object – first with one eye closed, then with the other eye closed. You will see two very slightly different views of the object. Signals about these views go from your eyes to your brain. When you use both eyes, your brain combines the signals from these views, and you perceive a solid object.

Look at the pictures below, both at the same time, one with each eye. (You may find it easier to do this if you look at the pictures through two tubes, one held to each eye, as shown.)

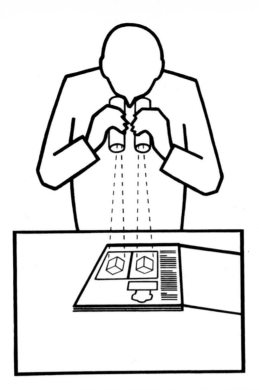

These pictures are the two slightly different views your eyes might have if you were looking at a cube. Images of these two pictures are formed in your eyes. Given time, your brain will combine the signals from these images, and you will see a solid cube. (You will need to concentrate, and you may have to look at the pictures for quite a long time before your brain combines the two images.)

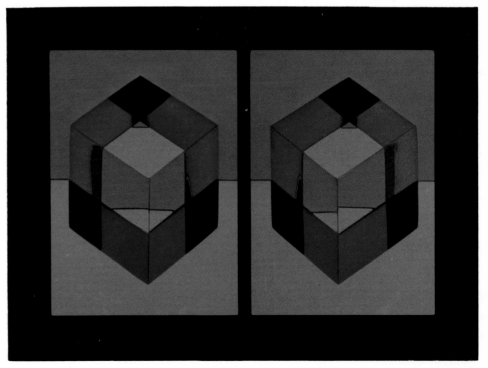

Knowing where

You use many clues to tell you about the position of objects. Which of these two cubes is in front of the other? How do you know?

You have to learn to use clues like this.

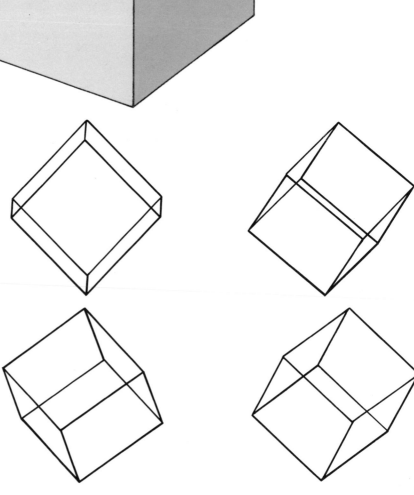

Knowing what

What do these diagrams represent?

Recognition plays an important part in your perception. You have learnt to recognize a cube from any position. You learn to perceive the world about you . . .

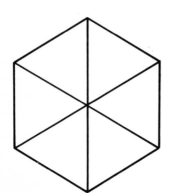

ow would you expect these cubes
feel?

he information you receive from
he sense may be strongly
ssociated with the information
u receive from your other senses.

So you may not have to touch an
object to know what it feels like, or
how heavy it is.

This strong association of
characteristics enables you to put
things into groups.

ook at the pictures on the opposite
age. They all represent cubes.
Vhen you see a cube you have not
een before, you still know what it
. You *classify* it with other cubes
nd expect it to be like a cube. With
he benefit of your previous
erceptions, you can cope with new
bjects. Of course, you may have to
odify your perception of an
bject when you discover more
bout it . . .

What is this?

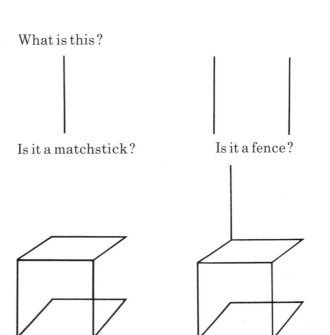

Is it a matchstick?   Is it a fence?   Is it a cube?

Yes, it is a cube . . .   No, it is not a cube . . .   It is a chair.

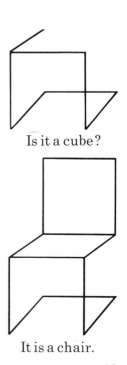

Perception

Perceiving things

Look at this photograph.

What do you see?

Stick insects look so like the bits of twig they live on that you may have to really look for them in order to see them. But your brain is receiving the same information from your eyes, whether you perceive the stick insects or not. Perception goes beyond the information given . . .

Look at this drawing. What does it represent? Is it a duck? Or is it a rabbit? It is ambiguous. You can perceive it in two different ways. In other words, you have two different *hypotheses* for what it represents, and your brain cannot choose between them.

Perceptions may be regarded as hypotheses based on information both from your senses and from your memory store. When there is not enough information, several hypotheses are possible. And your perception can change even when the information from your senses stays the same.

The impossible triangle

Look at this triangle.

Each corner looks correct. But surely you could not have these views of the corners if you were looking at a real triangle . . . The triangle looks quite wrong. But the image in your eyes so clearly suggests a triangle that your brain accepts this hypothesis. And so you see a triangle.

The puzzle

Look at this wall of black and white tiles.

Although the rows of tiles are quite straight, they look as if they are not. No one really understands why this illusion occurs. Solving puzzles like this can help us to understand more about perception.

Perception in everyday life

The moon illusion

Have you ever noticed that the moon looks much larger when it is low in the sky?

This is an illusion. When the moon is high in the sky, you have no clue as to how far away it is. When the moon is low in the sky, the ground gives you information about distance. You use this information to assess the size of the image of the moon in your eyes. And the moon seems larger.

Perception or imagination?

There is another way of looking at the moon. With a little bit of *imagination*, you can see the Man in the Moon. You see him, even though he is not there. The moon has misleading shadows which suggest a face.

Perception and imagination are closely linked. Perception involves making sense of very little information. This needs imagination.

Imaginative thinking is perception which is not controlled by the senses. You use imaginative thinking when you see faces in clouds, in bonfires – or even in tea-leaves.

Filtering out unwanted information

How do you cope with all the sensory information that constantly surrounds you?

You are surrounded by sounds of all kinds. You attend to some of these, but not to others. When several people are talking at once, you can attend to what one person is saying if that person has a distinctive voice. For example, you can usually follow a female voice in a room full of male voices.

It is the same with visual information. You pay attention to the words on the page you are reading. But you probably pay very little attention to the decoration down the side of the page, or to spelling misstakes, until your attention is drawn to them.

Perception of other people

Look at the people around you.

What do you notice about them?

3 Bob is now in conversation with Sue.

1 Sue and Peter are talking. Bob comes in.

4 Peter gets up to go.

2 Bob sits down.

5 Peter goes out.

Does Sue like talking to Peter?
Is Peter relaxed?

You perceive a lot about other people, just from the way they smile or frown, or from the way they hold themselves. But there are other clues that are not quite so obvious...

Is Sue interested in what Bob is saying?

Often without being fully aware of doing so, you respond to another person's interest – you look more closely, and the pupils of your eyes may get bigger.

How is Peter feeling?

When people are tense, it is reflected in the way they hold themselves, especially in the way they hold their hands, or cross and uncross their legs.

You learn to recognize these clues. And the way people look often influences the way you expect them to behave. This plays an important part in your perception of other people...

Would you ask Peter to a party?
Would you ask Bob to look after your money?

Our changing view of the world

Seeing faces

You have been looking at people all your life, and your ability to recognize a face is very strong. You do not need many clues. And you know that a face has a back, a top and sides. But you may not always have seen faces this way ...

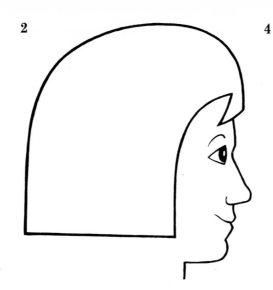

2

When you were a baby, you might not have recognized this as a face, because part of it is hidden.

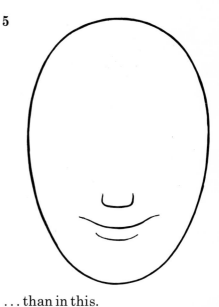

3

Even when you were three months old, this pattern would have interested you more than any other.

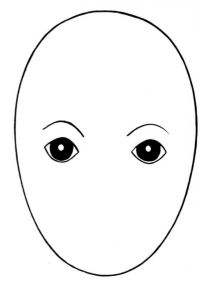

4

And when you were first born, you would have been more interested in this ...

5

... than in this.

You were born with some ability to perceive faces. But you had to learn to see faces the way you see them now.

Seeing patterns

Look at these dots.
Do they seem to form patterns?

The dots are equally spaced, yet your brain tends to group them in various ways. In the same way, you see patterns in the stars.

You were born with this tendency to group things into patterns. The patterns reflect the way your brain accepts the sensory information from which perceptions are derived.

Understanding the world about us

These abilities that we are born with – to perceive faces, to group things into patterns, and many others – form the basis of our developing perception . . . and our understanding of the world.

Chapter 12

How do we come to understand the world about us?

We are born with some ability to see, hear, smell, taste and feel. We are also capable of some simple actions. With these, we actively explore our world.

Exploration and discovery

Actions

At birth, we can do things like turn our heads, suck and grip. These *actions* are essential for our daily survival – they are also vital to the development of understanding.

We use our first actions to *explore* our surroundings. We suck and grasp at anything within reach, and soon come to recognize differences between objects.

By constant interaction with our surroundings, we discover more and more about the things around us, and how they are related to each other and to ourselves.

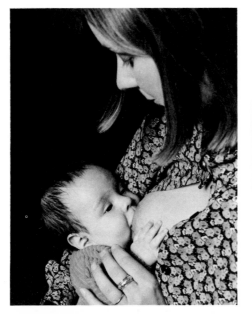

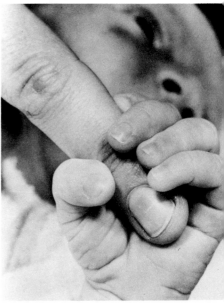

Means and ends

One of the things we discover is that our actions have certain results. The first time we shake a rattle, it is probably by chance. But we gradually learn how to make the noise on purpose.

As our knowledge of cause and effect increases, we are able to use intelligent behaviour to get what we want – we use a string to pull a toy towards us, and we climb on a chair to reach the table.

xpectations

ome scientists think that we are
so born with a few simple
pectations. It is not easy to find
ut what these expectations are,
ut they might include anticipating
at we can touch something that is
ithin reach.

xpectations and actions are
obably both vital to the
evelopment of understanding.
When our expectations are n
fulfilled, we use our actio
to explore the situatio
he experience we gain from this
lows us to modify our
expectation

a situation is completely
fferent from anything we have
ver met before, we may not be able
modify our existing ideas to
ccount for it. And, if our
xpectations are completely
lfilled, we discover nothing new.

ut, if the situation is only slightly
fferent from what we expected,
e can cope with it – we can modify
ur ideas and gain new
nderstanding.

any scientists think that this is
ow our understanding develops –
ep by step, through countless
teractions between our actions
nd our expectations.

What are her expectations?

Out of sight, out of mind?

Even when we can no longer see familiar objects, we usually behave as if they still exist.

But, when we are first born, we do not. We behave as if an object ceases to exist when it is out of sight.

As we grow older, we slowly discover that objects have *permanence* – this is one of the major achievements of infancy.

1 By the time a baby is old enough to reach out and pick up a toy, he will do so – even when it is partly hidden.

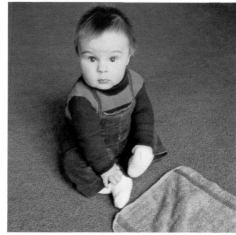

2 But, if the toy is completely hidden, he behaves as if it no longer exists – even if he watched as it was being hidden.

Beginning to think

By the time we are two years old, we are beginning to understand things in a mental, not just in a physical way. We are beginning to *think*. Thinking involves representing objects and events in our minds . . .

3 When he is older, he has no difficulty in finding a toy hidden in this way.

Symbols and signs

When we are about two years old, we begin to use *symbols* and *signs* to *represent* things. Dolls may symbolize people, bricks stand for cars, drawings portray daily events and words signify objects and activities.

This is a very important step. We use these symbols and signs to escape from reality. They allow us to recall the past, to contemplate the future and, eventually, to imagine the results of different actions – all in our minds.

Imitation

Imitation is probably the basis of our ability to use symbols and signs.

We begin to imitate at a few months old. By the time we are a year old, we can probably imitate many simple actions.

At first we only imitate actions – like clapping hands – that involve parts of our bodies we can see. Later we can imitate actions – like shutting our eyes – that involve our faces.

The little girl, on the right, is 'making up' for the first time. She is imitating something she saw her mother do earlier. This is an important step, for it must somehow involve a *mental image* of her mother's action.

Mental image

Representing actions and events by *mental images* is an essential part of thinking.

You use mental images in many ways. For example, suppose you decide to rearrange your bedroom. Before actually moving any of the furniture, you will probably try to *imagine* what it would look like in different positions . . .

We begin to form mental images of objects and events when we are about two years old. But it is several years before we can use mental images to anticipate changes or sequences of events.

The curved line
Here is a mental image game, that young children find difficult, if not impossible, to do.

> If this line were straightened, where would the ends come to?

Most young children seem to think the ends would stay in the same place. And, even if they are shown the line curved and then straightened, they cannot imagine the stages in between.

Answer *The ends of the line would reach the sides of the box.*

The graphic image

Drawing is great fun, but it also involves a serious attempt to represent what we know. Our ability to draw develops through a series of stages, and seems to reflect the growth of understanding.

3 As our understanding grows, we tend to draw all we know about an object – even the parts we cannot really see.

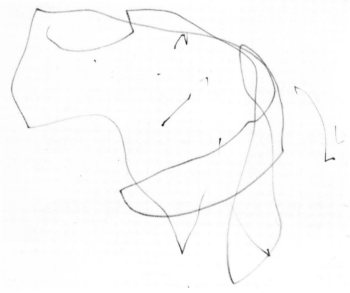

1 Our first drawings of objects look more like scribbles.

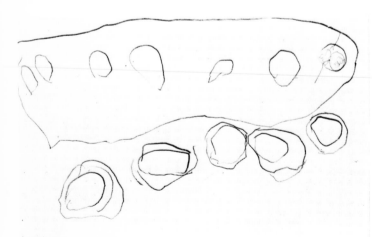

2 Later, the different parts of an object are recognizable – but we often fail to organize them properly.

4 Eventually, we take account of perspective – and draw conventional images.

Play

When we are children, *play* helps us escape from the restrictions and complexities of a real world we do not yet understand.

Play involves taking part in games, practising skills, making things and make-believe.

In games of make-believe, we find ways of our own to express the world around us. We use toys and everyday objects to represent all kinds of things . . .

Taking part in games

Practising skills

Making things

Make-believe

Language

Language is our most important system for representing things. It helps us to think, and it allows us to communicate with each other.

Learning to speak is a remarkable and rather mysterious process that scientists still don't fully understand. We receive no special training or instruction – we are simply exposed to language, and we learn to speak.

By the age of four, we have acquired most of the basic rules of grammar, and a vocabulary of more than a thousand words.

And we don't just imitate what we hear. From a very early age, we make up sentences of our own. Sentences we have never heard before. Sentences no one has ever heard before.

Language can be amazingly creative – as you will discover for yourself if you play the game on the next page.

The tharks have landed

Where have they landed?
Are they blue?

The four words in the title are enough to tell us which of the strange creatures is a thark. But what is a naggle? How many creatures are zibbing?

Words are socially agreed signs representing things and events. To use new words like 'naggle' or 'zib', we just have to agree on what they stand for.

The game
Try to describe the space scene using only the words listed.

Play the game with a friend, taking turns to see what you can tell each other. You will probably be surprised at how much you can say with the few words provided . . .

1 the/there
2 why/how
3 where/what
4 it/they
5 yes/no
6 big/small
7 blue/silver
8 red/yellow
9 one/many
10 square/round
11 thark/tharks
12 naggle/naggles
13 trantling/trantlings
14 wirv/wirvs
15 thing/things
16 has/have
17 is/are
18 come/came
19 can/cannot
20 do/doing
21 landed/landing
22 fly/flying
23 zibbed/zibbing
24 made/making
25 go/going
26 on/in
27 of/to
28 behind/from
29 up/away
30 and/with
31 moon/moons
32 creature/creatures
33 machine/machines
34 body/bodies
35 space/ground

The growth of thinking

As we grow up, our thinking passes through many characteristic stages. Each stage builds on the previous one as we gradually develop a more and more reliable understanding of the world around us.

Take *numbers* for example. How do we come to understand numbers? We have all learnt to count, but the understanding of numbers involves more than just learning to count.

Look at these dots.

Most five year olds will agree that there are as many red dots as there are blue ones. But, if the bottom row is spaced out . . .

They will say there are more blue ones! Few eight year olds will make this mistake.

Why do young children make mistakes like this? One theory is that they have not yet discovered a reliable method for comparing relative numbers. Even when they can count, they often use unreliable methods like length or size – and so make mistakes.

The development of thinking is a very slow process. One by one, we gradually come to understand ideas like number, length, weight and volume. It takes years to understand all these different ideas. And it is a very long time before we can grasp more complex ideas like *density* and why some things float and others sink . . .

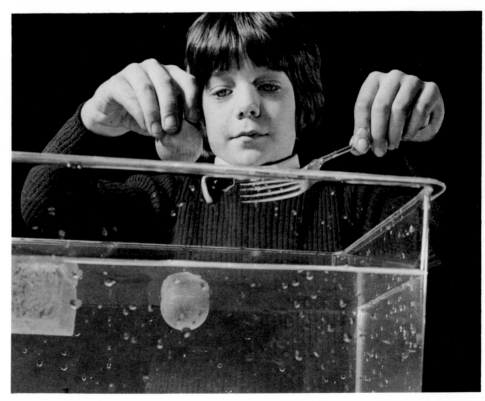

'Why did the fork sink?'
'Because it is heavy.'
'And why does the ball float?'
'Because it is light.'
'What about a large piece of wood and a small stone?'
'It's a heavy bit of wood, so perhaps it'll sink. I'll try . . . Well, it seemed heavier than the stone, but it can't really be as heavy . . . or is it that the stone and fork sank because they are heavy for such little things?'

115

Throughout childhood, our
thinking gradually changes as, one
by one, we overcome a number of
basic difficulties like this one . . .

'Have you got a brother?'
'Yes.'

'What's his name?'
'Jim.'

'Has Jim got a brother?'
'No.'

This little boy's answers reveal the
narrowness of our first outlook on
life. He can see only one
relationship – his brother's
relationship to himself. He cannot
see it from his brother's point of
view.

Before we can understand things
properly, we must be able to
imagine viewpoints other than
our own.

The growth of reason

As children, our thinking is restricted, and we find it difficult to solve problems without examining real objects. During adolescence, we develop new mental abilities that free us from this restriction and allow us to reason logically about possibilities – about what might be.

Try solving these puzzles. They may help you recognize a few of your own logical abilities.

The workers
Mike has less work to do than Alex.
Mike has more work to do than Dave.
Who has most work to do?

The race
The snail went faster than the dog.
The snail went slower than the frog.
The dog went faster than the cat.
In what order did they finish the race?

3 John's four journeys
These pictures represent four journeys. For each journey, the upper picture shows where John went, and the lower picture shows how he got there.

Describing these four journeys, John says 'Every time I went to York, I went by train.' Which *two* additional pictures must you see in order to find out whether John is lying or telling the truth?

Journey 1

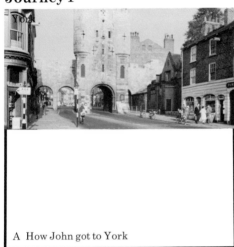

A How John got to York

Journey 2

B How John got to Edinburgh

Journey 3

C Where John went in this train

Journey 4

D Where John went in this car

Answers

1 The workers

'Mike has less work to do than Alex' – so Alex must have more work to do than Mike.

'Mike has more work to do than Dave' – so *Alex* must have the most work to do.

2 The race

'The snail went slower than the frog' – so the frog went faster than the snail.

'The snail went faster than the dog' and 'The dog went faster than the cat'

– so the frog came first
the snail came second
the dog came third
the cat came last.

These two puzzles involve drawing conclusions about relationships described only in words. Before adolescence, we cannot solve such problems in our heads. We need to refer to objects or drawings.

John's four journeys

You only need to see pictures A and D.

Journey 1

York

A Obviously you must find out if John was telling the truth about how he got to York.

Journey 2

Edinburgh

B How John got to Edinburgh

B This is irrelevant.

Journey 3

C Where John went in this train

C This is also irrelevant. John could have gone *anywhere* by train without affecting the truth of his statement.

Journey 4

D York

You must find out where John went by car because, if he *was* telling the truth, he cannot have gone to York by car. But he did – so he was lying.

This game demonstrates the importance of *negative* evidence in logical thinking. When testing the truth of a problem, we often make the mistake of looking only for information that seems to *confirm* the statement. This is why we are tempted to think that Journey 3 is significant.

How do we come to understand the world about us?

Expectation, action, symbols, thought, language and reason . . . as long as we respond to these, our understanding will continue to grow . . .

Index